FLORA OF TROPICAL EAST AFRICA

AMARANTHACEAE

C.C. TOWNSEND

H. Schinz in E. & P., Pf. ed. 2, 16C: 7–85 (1934); A. Cavaco, "Les Amaranthaceae de l'Afrique au sud du tropique du Cancer et de Madagascar", Mém. Mus. Nat. Hist. Nat. Paris, sér. B, Botanique, t. 13 (1962)

Annual or perennial herbs or subshrubs, rarely lianes. Leaves simple, alternate or opposite, exstipulate, entire or almost so. Inflorescence a dense head, loose or spike-like thyrse, spike, raceme or panicle, basically cymose, bracteate; bracts hyaline to white or coloured, subtending 1 or more flowers. Flowers ⚥ or unisexual (plants dioecious or monoecious), mostly actinomorphic, usually bibracteolate, frequently in ultimate 3-flowered cymules; lateral flowers of such cymules sometimes modified into scales, spines, bristles, hairs or hooks. Perianth uniseriate, membranous to firm and finally ± indurate, usually falling with the ripe fruit included, tepals free or somewhat fused below, frequently ± pilose or lanate, green to white or variously coloured. Stamens as many as and opposite to the petals, rarely fewer; filaments free or commonly fused into a cup at the base, sometimes almost completely fused and 5-toothed at the apex with entire or deeply lobed teeth, some occasionally without anthers, sometimes alternating with variously shaped pseudostaminodes (see note below); anthers 1–2-locular. Ovary superior, 1-locular; ovules 1-many, erect to pendulous, placentation basal; style very short to long and slender; stigmas capitate to long and filiform. Fruit an irregularly rupturing or circumscissile capsule, rarely a berry or crustaceous, usually with thin membranous walls. Seeds round to lenticular or ovoid; embryo curved or circular, surrounding the ± copious endosperm.

A large and mainly tropical family of some 65 genera and over 1000 species, including many cosmopolitan weeds and a large number of xerophytic plants; some are locally important vegetable or grain plants, or decoratives. General accounts of the family may be found by H. Schinz in E. & P. Pf., ed. 2, 16C: 7–85 (1934) and A. Cavaco, 'Les Amaranthaceae de l'Afrique au sud du tropique du Cancer et de Madagascar' in Mém. Mus. Nat. Hist. Nat. Paris, sér. B, Botanique 13 (1962).

The appendages found between the filaments in this family have been variously called staminodes or pseudostaminodes. The latter term, though cumbersome, has been preferred. The normal stamen complement in the *Amaranthaceae* is 5 – rarely less and never more. Thus, the intra-staminal appendages are always supernumerary to the stamens; they are not modified, sterile stamens as is understood by the term "staminode". The term "pseudostaminode" has also been standard in the family since its use by Schinz in his accounts of the family in both editions of "Die natürlichen Pflanzenfamilien".

Schinz divided the *Amaranthaceae* into two subfamilies – the *Amaranthoideae* (represented chiefly in the Old World) with anthers "4-facherig", and the *Gomphrenoideae* (represented chiefly in the New World) with anthers "2-facherig". It seems to me that Hutchinson, Families of Flowering Plants ed. 3: 545 (1973), is more correct in describing the anthers in the family as "1- or 2-locular". Apart from this, the arrangement used by Schinz is broadly employed in the present account, the genera in the F.T.E.A. region being deployed as follows:–

Subfamily *Amaranthoideae* : Anthers bilocular.

Tribe *Celosieae*: Ovules few to numerous, very rarely solitary and then not constantly so in the species showing this character. *Celosia, Hermbstaedtia.*

Tribe *Amarantheae*: Ovules constantly solitary.

Subtribe *Amaranthinae*: Seed erect; radicle downwardly directed. *Amaranthus, Digera, Neocentema.*

Subtribe *Achyranthinae*: Seed pendulous; radicle upwardly directed. *Sericostachys, Sericocomopsis, Centemopsis, Lopriorea, Cyathula, Allmaniopsis, Pupalia, Dasysphaera, Volkensinia, Aerva, Nothosaerva, Psilotrichum, Achyranthes, Centrostachys, Achyropsis, Pandiaka.*

Subfamily *Gomphrenoideae*: Anthers unilocular.

Tribe *Brayulineae*: Flowers solitary or fasciculate in the leaf axils; stamens (in the E. African plant) perigynous. *Guilleminea*

Tribe *Gomphreneae*: Flowers in spikes or capitula, axillary or terminal; stamens always hypogynous.

Subtribe *Froelichiinae*: Stigma capitate, penicillate or depressed and shortly bilobed, never with subulate branches. Flowers never compressed. *Alternanthera.*

Subtribe *Gomphreninae*: Stigma-branches subulate or distinctly bilobed, if stigma capitate then the flowers compressed. *Gomphrena, Iresine.*

During the course of studies in African Amaranthaceae, pollen morphology has proved very useful in determining generic affinity, and a general pollen survey would undoubtedly shed a good deal of light on the broad classification of the family. Investigation has not yet, however, been on a sufficiently wide basis to attempt such a revised arrangement.

SIMPLE KEY TO GENERA BASED ON READILY OBSERVED CHARACTERS

This key should be used with caution. If in doubt, the second key, p. 5, based on more technical characters should be employed.

Bracteoles with a projecting dorsal keel along at least the upper part of the midrib; inflorescences subglobose or shortly cylindrical, sessile above a pair of leaves ... 24. **Gomphrena**

Bracteoles never keeled; inflorescences very rarely sessile above a pair of leaves:

- Leaves alternate:
  - Inflorescences globose, sessile, axillary, flower parts with long, squarrose awns; low subshrub of northern Kenya .... 11. **Allmaniopsis**
  - Inflorescences not globose, sessile and axillary, not with long, squarrose awns:
    - Procumbent plant with rather succulent, spathulate-obovate leaves; very local on seashore, or shore of lakes in the coastal belt .... 2. **Hermbstaedtia**
    - Plant not having succulent, spathulate-obovate leaves:
      - Plant with all the flowers fertile, bearing stamens and/or ovary and style:
        - Flowers densely white-woolly:
          - Leaves usually obviously pilose at least on the lower surface (glabrous in some forms from Zanzibar and the coast, where *Nothosaerva* is not recorded) .... 15. **Aerva**
          - Leaves glabrous or with a few fine hairs only 16. **Nothosaerva**
        - Flowers not densely white-woolly, tepals glabrous or furnished with a few multicellular hairs towards the base:

Flowers unisexual, ♂ near the top of the inflorescence or scattered among ♀; flowers green, never tinged with white or pink and never with more than a single, central nerve .... 3. **Amaranthus**

Flowers always ⚥; tepals in several species tinged with or wholly white or pink when fresh, and in several with 3 or more nerves .... 1. **Celosia**

Plant with modified sterile flowers present (frequently one on each side of a fertile flower), with the appearance of scales, spines or wings:

Each bract with a single fertile flower subtended on each side by a modified sterile flower 4. **Digera**

At least some bracts subtending 2–3 fertile flowers, which with the modified flowers form a burr-like fruiting unit containing 2–3 seeds .... 5. **Neocentema**

Leaves opposite, at least the majority (scattered alternate leaves and branches may sometimes occur in *Psilotrichum*):

Modified sterile flowers (formed of hairs, spines, bristles or scales) present, frequently one on each side of a fertile flower:

Sterile flowers of greatly elongating plumose hairs; a tall forest climber .... 6. **Sericostachys**

Sterile flowers of bristles, scales or spines, never of hairs; sometimes scrambling but not tall-climbing:

Sterile flowers of fine, straight, flexible bristles:

Leaves cordate or truncate at the base, with an abrupt, short petiole; flowers green or reddish green in life .... 13. **Dasysphaera**

Leaves shortly or long-cuneate into a long petiole; flowers (at least the stamens and style, and often the inner surface of the tepals) bright magenta .... 14. **Volkensinia**

Sterile flowers of glochidiate, hooked or straight but rigid spines:

Spines of sterile flowers straight, neither glochidiate nor hooked .... 10. **Cyathula**

Spines of sterile flowers glochidiate or uncinately hooked:

Spines of sterile flowers in 3 or more rigidly stalked clusters, stellately divergent from the top of the stalk in fruit..... 12. **Pupalia**

Spines of sterile flowers separate to the base, or meeting in a very short, weak stalk 10. **Cyathula**

Modified sterile flowers absent, all flowers fertile:

Flowers unisexual on separate plants, minute (tepals 1–1.25 mm.), in terminal panicles; succulent cultivated plant with variegated, usually retuse leaves .... 25. **Iresine**

Flowers hermaphrodite; plant without the above combination of characters:

Inflorescences sessile and axillary; tepals at least 1.5 mm.; stamens 5 (2 filaments sometimes anantherous):
  Tepals densely woolly, 1-nerved, fused to ± halfway .... 22. **Guilleminia**
  Tepals glabrous and 1-nerved or barbellate-pilose and 3-nerved, free .... 23. **Alternanthera**
Inflorescences not sessile and axillary, or if so (*Nothosaerva*) tepals ± 1.25 mm. and stamens only 2:
  Flowers in part or all of the inflorescence 2-several per bract:
    Leaves auriculate-amplexicaul; pseudostaminodes absent.................. 9. **Lopriorea**
    Leaves not auriculate-amplexicaul; pseudostaminodes present:
      Annual (? sometimes short-lived perennial) herb with filiform to lanceolate-elliptic leaves; inflorescence commonly red.. 8. **Centemopsis kirkii**
      Small bushy shrubs with broader leaves; inflorescence white to greyish or stramineous .... 7. **Sericocomopsis**
  Flowers solitary within each bract:
    Robust aquatic or marsh perennial; perianth 6–8 mm.; upper tepal in flower slightly longer than the remainder, with a sharper and often slightly recurved tip 19. **Centrostachys**
    Plant rarely aquatic or paludal, if so the perianth not exceeding 2.5 mm.; upper tepal not differentiated:
      Fruits distinctly deflexed, to at least 45°:
        Tepals densely hairy, prominently nerved 21. **Pandiaka lanuginosa**
        Tepals glabrous, very obscurely nerved 18. **Achyranthes**
      Fruits not distinctly deflexed to at least 45°:
        Flowers very small, tepals ± 1.25 mm., villous dorsally; leaves not linear-filiform .... 16. **Nothosaerva**
        Flowers not so small, if tepals less than 2 mm. then they are glabrous or the plant has linear-filiform leaves:
          Inflorescences fasciculate, several of the ultimate racemes meeting at a single point and sessile:
            Leaves linear to very narrowly elliptic, glabrous or thinly puberulent .... 8. **Centemopsis fastigiata**
            Leaves oblong-ovate to oblong-elliptic, densely pilose especially on the lower surface ... 18. **Achyranthes fasciculata**
          Inflorescences not fasciculate, the ultimate racemes not thus condensed:
            Tepals ± 1.5 mm., deeply sulcate below between the midrib and the two broad, blunt, lateral nerves .... 8. **Centemopsis filiformis**

Tepals, if less than 2 mm., not deeply bisulcate below:
  Tepals glabrous:
    Tepals with a single distinct or obscure midrib only..... 20. **Achryopsis** (part)
    Tepals strongly to feebly 3-nerved, at least basally:
      Tepals cucullate at the apex, the midrib not excurrent in a mucro .... 20. **Achryopsis** (part)
      Tepals not cucullate at the apex, the midrib excurrent in a mucro........... 8. **Centemopsis** (part)
  Tepals pilose:
    Inflorescence-branches slender, width with flowers rarely exceeding 1 cm. and if so (*P. majus*) then tepals finely multinervose with short, fine, appressed hairs .... 17. **Psilotrichum**
    Inflorescence-branches stouter, width exceeding 1 cm., tepals 3–5(–7)-nerved with long, ± spreading hairs..... 21. **Pandiaka**

KEY TO GENERA BASED ON MORE TECHNICAL CHARACTERS

Bracteoles with a projecting dorsal keel along at least the upper part of the midrib .... 24. **Gomphrena**
Bracteoles with no dorsal keel on the midrib:
  Stamens fused into a tube, free portions of filaments very short, anthers alternating with conspicuous pseudo-staminodes .... 2. **Hermbstaedtia**
  Stamens not fused into a tube, the filaments free for at least half their length:
    Flowers unisexual, ♂ situated towards the top of the inflorescences or scattered among ♀ .... 3. **Amaranthus**
    Flowers ⚥, or rarely unisexual on separate plants:
      Modified sterile flowers (consisting of hairs, spines, bristles or scales) present, frequently one on each side of each fertile flower:
        Sterile flowers of greatly elongating plumose hairs; tall climber to about 6 m. or more.......... 6. **Sericostachys**
        Sterile flowers of hooked or straight spines, bristles, bracteoliform processes or scales, never of hairs; not tall climbers, though sometimes scrambling to ± 2 m.:
          Leaves opposite:
            Pseudostaminodes absent; filaments linear and free almost or quite to the base:
              Sterile flowers formed of uncinately hooked spines .... 12. **Pupalia**
              Sterile flowers formed of straight, fine bristles .... 13. **Dasysphaera**

Pseudostaminodes present between the filaments and ± fused to them:
Sterile flowers of fine, straight bristles 1–2 cm. long; flowers bright red .... 14. **Volkensinia**
Sterile flowers of uncinately hooked or short, straight, rigid spines; flowers not red .. 10. **Cyathula**
Leaves alternate:
Inflorescences sessile, globose, axillary; sterile flowers of long-aristate bracteoliform processes; low subshrub 11. **Allmaniopsis**
Inflorescences spicate, pedunculate; sterile flowers of accrescent antler- or wing-like scales; herbs:
Fruiting partial inflorescence a burr-like unit containing several seeds; ovary compressed, the apex with a transverse thickened rim sloping away from each side of the style; at least some bracts subtending more than one flower .... 5. **Neocentema**
Fruiting partial inflorescences of a single fruit subtended by two pairs of bracteolate scales or wings; ovary scarcely compressed, the apex with a circumferential rim; flowers constantly solitary in the axils of bracts .... 4. **Digera**
Modified sterile flowers lacking, only fertile present:
Stamens 1–2; flowers minute .... 16. **Nothosaerva**
Stamens 4–5; flowers minute to larger:
Leaves alternate:
Pseudostaminodes present; ovary with a solitary ovule .... 15. **Aerva**
Pseudostaminodes absent; ovary usually with several ovules, 1–2 only in *C. stuhlmanniana* .... 1. **Celosia**
Leaves opposite:
Flowers unisexual on separate plants (succulent cultivated plant) .... 25. **Iresine**
Flowers ⚥
Robust aquatic or marsh plant; perianth 6–8 mm.; upper tepal longer and narrower than the remainder in flower .... 19. **Centrostachys**
Plant rarely aquatic or paludal, if so the perianth not exceeding 2.5 mm.; upper tepal not narrower and longer than the remainder in flower:
Inflorescences all sessile and axillary:
Tepals densely woolly, sinuose-lanuginose, fused to about half-way, lobes mostly hyaline and delicate .... 22. **Guilleminia**
Tepals free, firm .... 23. **Alternanthera**
Inflorescences not all sessile and axillary:
Inflorescences with more than one flower within at least some bracts:
Leaves auriculate-amplexicaul; pseudo-

staminodes absent .... 9. **Lopriorea**
Leaves not auriculate-amplexicaul; pseudostaminodes present:
Plants with broad, not linear leaves; inflorescence not red .... 7. **Sericocomopsis**
Plants with linear leaves; inflorescence usually red .... 8. **Centemopsis**
Inflorescences with only one flower within each bract:
Pseudostaminodes absent or (very rarely) shortly deltoid and entire 17. **Psilotrichum**
Pseudostaminodes present, dentate or with a fimbriate dorsal scale:
Tepals pilose throughout .... 21. **Pandiaka**
Tepals glabrous, or floccose below the middle only:
Perianth sharply deflexed in fruit, or the inflorescence fasciculate and formed of several spikes of dissimilar length .... 18. **Achyranthes**
Perianth not sharply deflexed in fruit; inflorescence of single spikes:
Outer tepals with the apex±cucullate, the midrib not excurrent in a mucro .... 20. **Achyropsis**
Outer tepals with the apex not cucullate, shortly mucronate with the excurrent midrib:
Leaves linear to filiform, or the outer tepals at least 3.5 mm............... 8. **Centemopsis**
Leaves elliptic-ovate; outer tepals 3 mm .... 20. **Achyropsis**

1. CELOSIA

L., Sp. Pl.: 205 (1753) & Gen. Pl., ed. 5: 96 (1754); C.C. Townsend in Hook., Ic. Pl. 38(2) (1975)

Annual or perennial herbs, sometimes rather woody at the base, occasionally scandent. Leaves alternate, simple, entire or somewhat lobed. Flowers small, bibracteolate, ⚥ , in spikes or more generally dense to lax axillary and terminal bracteate thyrses (the uppermost frequently forming a panicle), lateral cymes lax to dense and forming sessile clusters, the inflorescence then spiciform. Perianth-segments 5, free, equal. Stamens 5, the filaments fused below into a short sheath, the free portions deltoid below and filiform above; anthers bilocular. Ovary with few-many (rarely solitary) ovules; style elongate to almost obsolete; stigmas usually 2–3. Capsule circumscissile, sometimes thickened above. Seeds black, usually strongly compressed and shining, roundish, feebly to strongly reticulate, grooved or tuberculate.

About 50 species, in the tropics and subtropics of both hemispheres.

The description above differs from the concept of Schinz in ed. 2 of the Pflanzenfamilien (1934) in referring to *Hermbstaedtia* the species placed by Schinz in the subgenera *Pseudohermbstaedtia* and *Gomphrohermbstaedtia* of *Celosia*.

Style 5–7 mm.; perianth-segments 6–10 mm. . . . . 12. *C. argentea*
Style less than 5 mm.; perianth-segments less than 6 mm.:
Testa of seeds almost smooth, with a very faint and fine reticulate pattern; areolae almost plane, rhomboidal to square centrally, becoming more elongate towards the margin:
Ovary (Fig. 1) becoming strongly spongy-incrassate at the apex in fruit, in the mature fruit the apex thicker than the lower part, the style set in the concave centre; ovules 2–4 . . . . 1. *C. anthelminthica*
Ovary only slightly thickened in fruit, at most truncate at the apex; ovules 5 or more:
Perianth-segments shortly mucronate with the excurrent midrib, pale, membranous with a narrow rather less translucent central vitta . . . . 4. *C. trigyna*
Perianth-segments not mucronate, the midrib ceasing below the apex, when dry with a white membranous border, usually dark brown or blackish centrally:
Upper spikes of flowers without or with very reduced leaves, forming a broad terminal panicle; style ± 0.5 mm., slightly shorter than or subequalling the stigmas . . . . 3. *C. polystachia*
Plant without a terminal panicle, the terminal spike with at most 1–2 slender branches below; style very short, much shorter than the stigmas . . . 2. *C. schweinfurthiana*
Testa of seeds strongly sculptured with grooves and/or punctate impressions, deeply reticulate with the areolae ± convex:
Lower leaves hastate, with broad and shallow to narrower and longer lobes:
Perianth-segments 2.5–2.75 mm.; spikes slender; ovules 2–9:
Perianth-segments 3–4(–5)-nerved; ovary 8–9-ovulate . . . . 5. *C. patentiloba*
Perianth-segments 1-nerved; ovary 2–3-ovulate . . . . 6. *C. fadenorum*
Perianth-segments 4–5 mm.; spikes dense, the axis very closely furnished with multicellular hairs; ovules 14 or more . . . . 9. *C. hastata*
Lower leaves not hastate:
Perianth-segments 3- or more nerved to the middle or above:
Perianth-segments 2–2.5 mm.; ovules 1–2 . . . . 7. *C. stuhlmanniana*
Perianth-segments 3–6 mm.; ovules 5–20 or more:
Inflorescence drying pale chestnut-brown; perianth-segments conspicuously 5–7-nerved, the laterals reaching nearly to the apex . . . . 9. *C. hastata*
Inflorescence drying silvery-white; perianth-segments 3-nerved, the laterals not reaching much above the middle . . . . 8. *C. isertii*
Perianth segments 1-nerved, or rather feebly 3-nerved with the lateral nerves not or scarcely reaching to the middle of each segment:
Inflorescences all ± globose, axillary, ± sessile or rarely on a peduncle up to 2 cm. long . . . . . . 11. *C. globosa* var. *globosa*
Only the lower inflorescences, if any, globose and ±

sessile, at least the terminal oblong to cylindrical or paniculate, pedunculate:
Perianth-segments with at least the margins silvery-white or hyaline when dry:
Perianth-segments (3–)3.5–5 mm.; spikes usually many-flowered and stout, up to 2 cm. wide; ovules more than 12 .... 8. *C. isertii*
Perianth-segments 2–3 mm.; spikes slender; ovules 8 or less .... 10. *C. elegantissima*
Perianth-segments in no part silvery-white or hyaline, drying buff-coloured, darkest in the centre .... 11. *C. globosa* var. *porphyrostachya*

1. **C. anthelminthica** *Aschers.* in Schweinf., Beitr. Fl. Aeth.: 176 (1867); P.O.A.C: 172 (1895); E.P.A.: 55 (1953); U.K.W.F.: 132 (1974); C.C. Townsend in Hook., Ic. Pl. 38 (2): 13, t. 3726 (1975). Type : Ethiopia, Dschadscha, *Schimper* 2173 (K, iso.!)

Much-branched scrambling or straggling, bushy herb, prostrate to erect or scandent, ± 0.3–3 m. and probably more. Stem, branches and inflorescence-axis ridged, glabrous to rather densely furnished with multicellular brownish hairs. Leaves cordate-ovate to deltoid-lanceolate, subacute to acuminate, glabrous or with scattered multicellular hairs on the veins of the lower leaf surface; lamina of leaves from central part of stem 1.5–8 × 1–5 cm., very shortly cuneate to truncate or cordate at the base, with a slender 1–4 cm. petiole; upper and branch leaves smaller, more shortly petiolate. Inflorescence of axillary and terminal elongate thyrses 8–40 cm. long, formed of distant few- to rather many-flowered cymose clusters which may be loose with slender branches up to ± 5 mm. long or dense and congested; clusters thus varying from ± 3–18 mm. in diameter. Bracts and bracteoles lanceolate-acuminate, 1–1.5 mm., glabrous, membranous with a brownish or yellowish midrib. Perianth-segments oblong, ± 2.5 mm., concave and strongly hooded at the apex, glabrous, 1-nerved, pink, white or greenish, drying straw-coloured with a darker or greenish central vitta. Free portion of filament longer than the sheath, sinuses slightly rounded with no intermediate teeth. Stigmas 2(–3), 0.75 mm., divergent. Style practically absent; ovary 2–4-ovulate, thickened above. Capsule oblong-clavate, exserted, 2–2.5 mm., the apex spongy-incrassate so that the style is frequently inserted in a definite depression. Seeds 1–1.25 mm., compressed, shining, black, ornamented with a very fine reticulate pattern which is faintest in the almost smooth centre of the seed. Fig. 1.

UGANDA. Karamoja District: Moroto, Sept. 1956, *J. Wilson* 252! & 15 July 1958, *Kerfoot* 406! & Lodoketeminit, 13 Jan. 1959, *Kerfoot* 976!
KENYA. Northern Frontier Province: Mt. Nyiru [Nyiro], July 1960, *Kerfoot* 2024!; Nyeri, *van Someren* 1707!; Masai District: Emali Hill, Mar. 1940, *V.G. van Someren* 172!; Teita District: Voi Park H.Q. Hill, Dec. 1966, *Greenway & Kanuri* 12823!
TANZANIA. Shinyanga, Feb. 1933, *Bax* 38!; Masai District: Lemuta, July 1962, *Oteke* 200!; Lushoto District: Mombo Forest Reserve, Sept. 1960, *Semsei* 3088!
DISTR. **U**1; **K**1–4, 6, 7; **T**1–3; Rwanda, Ethiopia
HAB. Frequently in shade, rambling over bushes in scrub or in relict forest, along streams, growing in clumps in grassland or sprawling over rocks; 500–2280 m.

SYN. *C. acroprosodes* Bak. & C.B.Cl. in F.T.A. 6(1): 22 (1909); Hochst. in Bot. Zeit. 14: 598 (1856), *nomen*; F.D.O.-A. 2: 210 (1932). Type: Ethiopia, Dschadscha, *Schimper* 2173 (K, lecto.!)
C. digyna Suesseng. var. *cordata* Suesseng. in Mitt. Bot. Staats., München 1: 74 (1951). Type: Tanzania, Moshi District, Chala Crater, *Geilinger* 4745 (K, holo.!)

2. **C. schweinfurthiana** *Schinz* in E.J. 21: 178 (1895); P.O.A.C: 172 (1895); F.T.A. 6(1): 22 (1909); F.D.O.-A. 2: 209 (1932); F.P.N.A. 1: 128

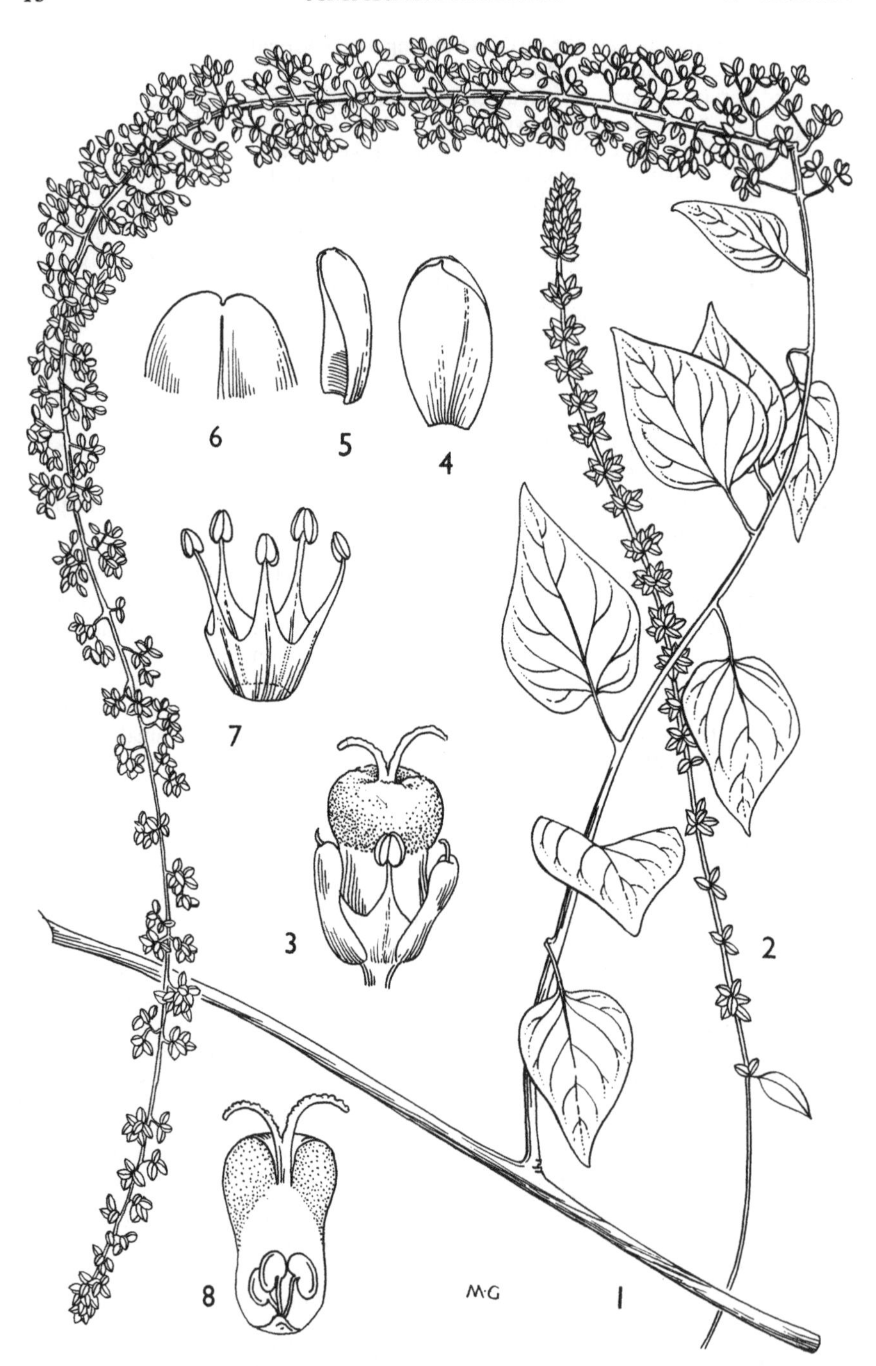

FIG. 1. *CELOSIA ANTHELMINTHICA*—**1, 2,** flowering branches, with lax and compacted cymes respectively, × $\frac{2}{3}$; **3,** fruit (one tepal removed), × 14; **4, 5,** tepals, front and side views respectively, × 16; **6,** tepal-apex, flattened, × 16; **7,** stamens, × 16; **8,** gynoecium, longitudinal section, × 16. 1, 3–8, from *Newbould* 5774; 2, from *Newbould* 6437. Drawn by Mary Grierson; reproduced from 'Hooker's Icones Plantarum'.

(1948); F.P.S. 1: 118 (1950); Hauman in F.C.B. 2: 18 (1951); E.P.A.: 56 (1953); U.K.W.F.: 132 (1974); C.C. Townsend in Hook, Ic. Pl. 38(2): 17, t. 3727 (1975). Type: Zaire, Naporruporru, *Schweinfurth* 3062 (K, isolecto.!)

Perennial herb, often suffrutescent, varying considerably in habit from a prostrate plant rooting at the nodes to an erect herb 15 cm. tall or a climber rambling to 5 m. Stem and branches usually strongly ridged or sulcate-striate, glabrous or sometimes with a few short multicellular hairs, particularly about the upper nodes. Leaves lanceolate to lanceolate- or deltoid-ovate, acute to acuminate, glabrous or more frequently with scattered, short, multicellular hairs on the lower basal surface; lamina of main stem leaves 2.2–10 × 1–7 cm., shortly cuneate to truncate or subcordate at the base, ± decurrent into a slender 1–4.5 cm. petiole; upper and branch leaves smaller and usually narrower. Inflorescence of axillary and terminal elongate thyrses, 7–42 cm. long and 0.5–2.8 cm. wide, 0–2-branched at the base and without any leafless terminal panicle, formed of ± 1–10-flowered, distant or more rarely approximate, dense to very lax (pedicels to ± 3 mm.) cymose clusters, 2–12 mm. in diameter. Bracts and bracteoles lanceolate or deltoid, 0.5–1.25 mm., glabrous, membranous with a brownish or blackish midrib. Perianth-segments oblong-elliptic, 2 mm., 1-nerved, obtuse and often hooded, pale greenish to white, in the dry state pale brownish to blackish with a narrow, pale, delicately scarious margin. Free portion of filaments longer than the sheath, sinuses shallowly rounded with no intermediate teeth. Stigmas 2, reflexed, much longer than the very short style; ovary (5–)6–8(–10)-ovulate. Capsule oblong, ± 3 × 1 mm., dark, usually distinctly exserted, ± truncate and faintly to distinctly rugose-incrassate at the apex. Seeds ± 1 mm., compressed, shining, black, ornamented with a very fine reticulate pattern.

UGANDA. Toro District: Bwamba, Nov. 1935, *A.S. Thomas* 1526!; Mbale District: Budadiri, Jan. 1932, *Chandler* 504!; Mengo District: Mukono, Oct. 1914, *Dummer* 296!
KENYA. N. Kavirondo District: Mlaba, June 1964, *Tweedie* 2832! & Kakamega Forest, Jan. 1976, *Sturrock* 515!; Kilifi District: Sabaki, 3 Nov. 1961, *Polhill & Paulo* 705!
TANZANIA. Moshi, Nov. 1955, *Milne-Redhead & Taylor* 7210!; Mpwapwa, Feb. 1966 *Leippert* 6290!; Morogoro District: Bunduki, Jan. 1935, *E.M. Bruce* 430!
DISTR. U2–4; **K**1, 3, 5, 7; **T**1–3, 5–7; Sudan, Ethiopia, Zaire, Angola
HAB. As ground cover along forest rides, margins and clearings, especially near water or scrambling in thicker forest, otherwise in roadside or coastal bushland, or as a weed of cultivation; 3–1550 m.

SYN. *C. intermedia* Hochst. in Bot. Zeit. 14: 598 (1856), *nomen*
C. oblongocarpa Schinz in Bull. Herb. Boiss. 4: 418 (1896). Type: Tanzania, Mpwapwa, *Stuhlmann* 251 (Z, holo.!)
C. macrocarpa Lopr. in E.J. 30: 6 (1901), *nom. illegit.*; F.T.A. 6(1): 21 (1909); F.D.O.-A. 2: 209 (1932). Based on a monstrosity, *Volkens* 2244 (Z!) from Tanzania, Kilimanjaro
C. schweinfurthiana Schinz var. *sansibariensis* Schinz in Bull. Herb. Boiss., sér. 2, 3: 9 (1903); F.T.A. 6(1): 22 (1909); F.D.O.-A 2: 209 (1932). Type: Tanzania, Dar es Salaam, *Hildebrandt* 1247 (K, isolecto.!)

3. **C. polystachia** *(Forssk.) C.C. Townsend* in Hook., Ic. Pl. 38(2): 23, t. 3728 (1975). Type: Yemen, Surdûd, 20 Feb. 1763, *Forsskal* (BM, holo.!)

Suffrutescent herb, bushy or scandent, ± 1–4 m. Stem and branches subterete, finely striate, glabrous or with a few very scattered multicellular hairs, for a variable length below the inflorescence with a glaucous waxy bloom. Leaves broadly lanceolate- to subcordate-ovate, acute or shortly acuminate, glabrous or with scattered short multicellular hairs on the lower surface, especially near the base; lamina of the leaves from the central part of the stem 2.5–8 × 1.8–4.5 cm., very shortly cuneate to truncate or subcordate at the base, with a 1.3 cm. petiole; upper and branch leaves smaller, often narrower, more shortly petiolate. Inflorescence of axillary simple or branched spike-like condensed thyrses ± 6–21 cm. long, formed of usually ± distant ± 1–8-flowered clusters 1–8 mm. in diameter, with a broad terminal leafless panicle up to ± 35 cm. long at the ends of the branches, formed of similar thyrses. Bracts and bracteoles broadly deltoid, acute, 1.25–2 mm., membranous with a dark vitta along the central nerve, ciliate with multicellular hairs. Perianth-segments elliptic-oblong, 2 mm., acute, subcucullate at the apex, 1-nerved with a dark vitta along the nerve, glabrous. Free portion of filaments much longer than the very short sheath, sinuses rounded with no intermediate teeth. Stigmas 2–3, longer than the style; ovary 2–4-ovulate [6–7(–11)-ovulate in one gathering from Somalia, described as var. *pluriovulata* Suesseng.]. Capsule ovoid, included, rounded and faintly rugose-incrassate at the apex. Seeds ± 1 mm., compressed, black, shining, with a very fine reticulate pattern.

KENYA. Northern Frontier Province: Dandu, Mar. 1952, *Gillett* 12649! & Moruethe, Mar. 1965, *Newbould* 7322!; Baringo District: Lake Baringo, Mar. 1901, *Johnston*!

DISTR. K1, 3; Sudan, Ethiopia, Somalia, SW. Arabia

HAB. In clay depressions, in deciduous bushland on rocky slopes and along banks of watercourses; 400–1060 m.

SYN. *Achyranthes polystachia* Forssk., Fl. Aegypt.-Arab.: 48 (1775)
Celosia populifolia Moq. in DC., Prodr. 13(2): 239 (1849); F.T.A. 6(1): 24 (1909); F.P.S. 1: 118 (1950); E.P.A.: 55 (1953). Type: Ethiopia, Modat, Aguar valley, *Schimper* II. 1038 (K, lecto.!)
[*Deeringia celosioides* sensu Bak. & C.B. Cl., F.T.A. 6(1): 16(1909), quoad pl. Johnstonii; Cavaco in Mém. Mus. Nat. Hist. Nat. Paris, sér. B, 13; 38(1962), *non* R. Br.]
Celosia populifolia Moq. var. *pluriovulata* Suesseng. in K.B. 4: 476 (1949); E.P.A.: 55 (1953). Type: Somalia, Mwred valley, *Glover & Gilliland* 727 (K, holo.!)

4. **C. trigyna** *L.*, Mant. Pl. Alt.: 212 (1771); P.O.A. C: 172 (1895); F.T.A. 6(1): 19 (1909); F.D.O.-A. 2: 208 (1932); F.P.N.A. 1: 126 (1948); F.P.S. 1: 117 (1950); Hauman in F.C.B. 2: 17 (1951); E.P.A.: 56 (1953); Cavaco in Mém. Mus. Nat. Hist. Nat. Paris, sér. B, 13: 44 (1962); F.P.U.: 102 (1962); U.K.W.F.: 132 (1974); C.C. Townsend in Hook., Ic. Pl. 38(2) 27, t. 3729 (1975). Type: Senegal, *Herb. Linnaeus* (S, holo., IDC microfiche 102. 19!)

Annual herb, erect, simple or branching from near the base upwards, (8–)30–120(–180) cm. Stem and branches green to reddish, sulcate or striate, glabrous or with subscabrid, short, few-celled hairs especially about the nodes. Leaves narrowly lanceolate to broadly ovate, acute to acuminate, glabrous or with short few-celled hairs on the lower surface about the base; lamina of the leaves from main stem (10–)20–85(–100) × (4–)10–40(–50) mm., subcordate to truncate or attenuate below, the lower margins often scabrid, ± decurrent along the slender, up to ± 5 cm. petiole; upper and branch leaves smaller and often narrower, more shortly stalked; all leaves often deciduous by the time of fruiting. Inflorescences axillary and terminal, simple or branched spike-like condensed thyrses ± 6.5–35 cm. long, formed of distant or (at least above) approximate few–many-flowered lax or congested and subglobose white or pinkish clusters 2–20(–30) mm. in diameter, in well-grown individuals the upper leaves much reduced so that a terminal panicle is formed; inflorescence-axis glabrous or sparingly furnished with multicellular hairs. Bracts and bracteoles ovate to oval-elliptic, ± 1.25–2 mm., scarious with a single nerve, margins minutely (often more coarsely at the base) erose-denticulate, glabrous. Perianth segments oval-elliptic, 1.75–2.75 mm., shortly mucronate with the percurrent single nerve, glabrous, scarious with a narrow less translucent vitta along the nerve, margins minutely denticulate, at least above. Free portion of filaments subequalling the sheath, sinuses rounded with no intermediate teeth; anthers red. Stigmas 2–3, longer than the very short style; ovary 6–8-ovulate. Capsule ovoid, 1.75–2.25 mm., included or a little exserted, rounded and not thickened at the apex. Seeds ± 0.75 mm., compressed, black, shining, with a rather fine reticulate pattern, the areolae being only very slightly convex.

UGANDA. Acholi District: Kitgum, Dec. 1931, *Hancock* in *A.D.* 2396!; Busoga District: Lolui I., May 1964, *G. Jackson* U. 129!; Mengo District: Entebbe, Apr. 1904, *E. Brown* 35!
KENYA. Meru, Aug. 1958, *Bogdan* 4627!; Kisumu, 22 Sept. 1915, *Dowson* 455!; Kwale, 1934, *McCraig* in *C.M.* 9170!
TANZANIA Ngara District: Keza, July 1960, *Tanner* 5041!; Lushoto District: Korogwe, Nov. 1962, *Archbold* 20!; Mbeya District: Igawa, Apr. 1962, *Polhill & Paulo* 1969!; Zanzibar I, Nov. 1873, *Hildebrandt* 1033!
DISTR. U1-4; K2, 4, 5, 7; T1-8; Z; practically throughout tropical Africa, also Namibia, South Africa, Madagascar, S. Arabia, Madeira; recently recorded as naturalised in Florida
HAB. Most frequently as a weed of abandoned or currently cultivated arable land, also in forest clearings, along woodland paths and roadsides, in short grassland or on damp ground by rivers; 1-1500 m.

SYN. *C. melanocarpos* Poir., Encycl. Méth. Suppl. 4: 318 (1816). Type: Senegal (P, holo.)
C. laxa Schumach. & Thonn., Beskr. Guin. Pl.: 141 (1827). Type: Ghana, Ada, *Thonning* 180 (C, holo.!)
C. triloba Meissn. in Hook., Lond. Journ. Bot. 2: 548 [misprinted 448] (1843). Type: South Africa, Natal, R. Umgani, *Krauss* 238 (K, holo.!)
C. trigyna L. var. *fasciculiflora* Moq. in DC., Prodr. 13(2): 241 (1849). Type: Sudan, Kordofan, *Kotschy* 285 (K, iso.!)
C. trigyna L. var. *adoensis* Moq. in DC., Prodr. 13(2): 241 (1849). Type: Ethiopia, Mt. Scholoda, *Schimper* I. 49 (K, iso.!)
C. minutiflora Bak. in K.B. 1897: 277 (1897); F.T.A. 6(1): 23 (1909). Type: Tanzania, Tabora District, Urambo, *Hannington* (K, holo.!)
C. semperflorens Bak. in K.B. 1897: 277 (1897). Type: Malawi, Blantyre, *Buchanan* 52 (K, holo.!)
C. digyna Suesseng. in Trans. Rhod. Sci. Assoc. 43: 8 (1951) & Mitt. Bot. Staats., München 1: 73 (1951) excl. vars. Type: Zimbabwe, Rusape, *Dehn* 1121/52 (K, iso.!)
C. trigyna L. var. *longistyla* Suesseng. in Mitt. Bot. Staats., München 1: 75 (1951), incl. subvars. Type: *Jacquin*, Pl. Hort. Bot. Vindob. Cent. 3, t. 15 (lecto.!)
C. trigyna L. var. *convexa* Suesseng. in Mitt. Bot. Staats., München 1: 75 (1951). Type: Mozambique, Namagoa, *Faulkner* K. 46 (K, lecto.!)
C. trigyna L. var. *brevifilamentosa* Suesseng. in Mitt. Bot. Staats., München 1: 75 (1951). Type: Kenya, Kilifi, *Jeffrey* 445 (K, lecto.!)

5. **C. patentiloba** *C.C. Townsend* in Hook., Ic. Pl. 38(2): 41, t. 3732 (1975). Type: Tanzania, Newala, *Hay* 61 (K, holo.!)

Erect annual herb up to ± 1.1 m., but generally much smaller, branched from the base. Stems ridged and striate, slender, sparingly furnished with short multicellular hairs. Lower leaves subglabrous, deltoid-pandurate with broad, blunt lateral lobes at the base, the lamina to ± 6 × 3.5 cm., the petiole 1.5–2 cm.; upper leaves lanceolate-hastate, rapidly reducing above and more shortly stalked, more obviously furnished with multicellular hairs. Inflorescences terminal and axillary, simple or the terminal slightly branched at the base, of elongate thyrses up to ± 18 × 0.6 cm., formed of distant sessile clusters (reduced cymes) 4–6 mm. in diameter. Bracts and bracteoles deltoid, acute, ± 1.5 mm., scarious with a percurrent yellowish midrib, usually ± ciliate below with multicellular hairs. Perianth-segments elliptic-oblong, 2.5–2.75 mm., glabrous or furnished with a few multicellular hairs on the lower dorsal surface, white (pale straw-coloured when dry), 3–4(–5)-nerved with the nerves more prominent at the base, minutely denticulate at the apex. Free portion of filaments subequalling or slightly shorter than the sheath, sinuses without teeth. Ovary 8–9-ovulate; stigmas 2(–3), longer than the ± 0.5 mm. style. Capsule ovoid, ± 2.5 mm., included. Seeds ± 1.2 mm., black, shining, compressed, with convex hexagonal or rhomboid areolae centrally and feebly punctate reticulating furrows in the marginal half.

TANZANIA. Newala, Apr. 1959, *Hay* 61!
DISTR. T8; known only from the type gathering
HAB. In semi- or full shade; 665 m.

6. **C. fadenorum** *C.C. Townsend* in Hook., Ic. Pl. 38(2): 122 (1975). Type: Kenya, Tana River District, Garissa–Thika, *R.B. & A.J. Faden* 74/798A (K, holo.!)

Erect perennial herb 60 cm. tall, branched in the lower half, glabrous throughout. Stems ridged and striate, slender. Lowest leaves deltoid-ovate, soon becoming deltoid-pandurate with broad blunt lateral lobes at the base, the lamina to ± 4.5×2.75 cm., the petiole 1.5–2 cm., upper leaves lanceolate-hastate with a narrower central lobe, the uppermost frequently linear with no basal lobes; small rounded and often falcate stipuliform leaves also subtending many stem and branch leaves. Inflorescence terminal, 7.5 × 0.8 cm., a spiciform thyrse formed of compacted cymes ± 0.5 cm. in diameter, only the lowest cymes being slightly distant. Bracts and bracteoles deltoid, very acute,±2 mm., white, scarious, glabrous, erose-denticulate above with an excurrent midrib. Perianth-segments elliptic-oblong, 2.5–3 mm., obtuse, glabrous, white, 1-nerved with the midrib excurrent in a very short mucro, erose-denticulate above. Free portion of filaments subequalling or slightly shorter than the sheath, sinuses rounded, without teeth. Ovary 2–3-ovulate; stigmas 2(–3), longer than the ± 0.5 mm. style. Capsule ± 1.75 mm., included. Seeds ± 1.2 mm., black, shining, compressed, with convex hexagonal or rhomboid central areolae and scarcely or very feebly punctate reticulating furrows around the margins.

KENYA. Northern Frontier Province: 6 km. N. of Uaso-Nyiro crossing on Dadaab–Wajir road, May 1977, *Gillett* 21247!; Tana River District: Garissa–Thika road 4.5 km. towards Thika from the turn-off to Galole, June 1974, *R.B. & A.J. Faden* 74/798A!
DISTR. **K**1, 7; not known elsewhere
HAB. Deciduous bushland; 260–280 m.

7. **C. stuhlmanniana** *Schinz* in Bull. Herb. Boiss. 4: 419 (1896); F.T.A. 6(1): 21 (1909); F.P.N.A. 1: 129 (1948); Hauman in F.C.B. 2: 25 (1951); Townsend in Hook., Ic. Pl. 38(2): 53, t. 3735 (1975). Type: Zaire, SW. of Lake Mobutu [Albert], *Stuhlmann* 3052 (Z, lecto.!)

Scrambling or straggling herb, or small liane, up to 6 m. in height. Stem and branches ridged, glabrous or sparingly furnished (often more densely so about the nodes) with multicellular brownish hairs. Leaves ovate-lanceolate to lanceolate or elliptic, acutely acuminate, glabrous or with scattered multicellular hairs on the veins of the lower leaf surface; lamina of leaves from central part of stem 3.5–15 × 1.5–8 cm., rather shortly cuneate or rounded below to a 1–3 cm. petiole, the upper leaves smaller, narrower and more shortly petiolate. Inflorescence of axillary simple or branched spike-like thyrses ± 8–25 cm. long, formed of distant few-flowered clusters and a broad leafless terminal panicle up to 70 cm. long and 30 cm. wide; inflorescence-axis and branches glabrous or commonly increasingly furnished upwards with multicellular hairs. Flower-clusters (condensed cymes) dense, 2–6 mm.in diameter. Bracts and bracteoles deltoid-ovate, ± 1 mm., acute, margins and usually at least the lower part of the single nerve with long multicellular hairs. Perianth-segments elliptic-oblong, 2–2.5 mm., acute, 3–4(–5)-nerved, moderately pilose with long multicellular hairs towards the base, scarious with a yellowish or buff centre. Free portion of filaments shorter than sheath; filaments at anthesis alternating with very short, bluntly deltoid teeth which soon disappear. Stigmas 2, longer than the style; ovary 1(2–)-ovulate. Capsule included, rounded-tapering into the style. Seeds 1–1.25 mm., compressed, black, shining, furnished with low rounded verrucae (convex areolae of the surface reticulation).

UGANDA. Toro District: Nyamwamba [Namwamba] Valley, Kilembe, Dec. 1934, *G. Taylor* 2567!; Ankole District: Near Kikagati, Sept. 1947, *Dale* U. 495!; Kigezi District: Amahinga, Mar. 1946, *Purseglove* 2017!

TANZANIA. Bukoba District: Karagwe, *Scott Elliot* 8139! & Kabirizi, Oct. 1931, *Haarer* 2218!; Kigoma District: Gombe Stream National Park, June 1970, *Clutton-Brock* 259A!
DISTR. U1, 2; T1, 4; Zaire, Rwanda, Zambia
HAB. In forest shade and thickets; 1200–1500 m.

8. **C. isertii** *C.C. Townsend* in Hook., Ic. Pl. 38(2): 57, t. 3736 (1975). Type: Ghana, Whydah Guineae, 1785, *Isert* (C, holo.!)

Perennial herb, often suffrutescent, varying considerably in habit from a lax creeping or decumbent plant to a scandent climber attaining 4 m. or probably more. Stem and branches ridged or striate, glabrous or thinly to (not in East Africa) rather densely furnished with brownish multicellular hairs. Leaves lanceolate-ovate to broadly or deltoid-ovate, shortly acuminate, glabrous or thinly furnished on the lower surface (especially along the veins) with multicellular hairs, occasionally a few hairs present on the upper surface also; lamina of main stem leaves 3.5–10 × 2.5–5 cm., shortly cuneate to truncate or subcordate at the base, rapidly narrowed and ± decurrent into the slender, 1.5–4 cm. petiole; upper and branch leaves smaller, often narrower and more attenuate to the base. Inflorescence of axillary and terminal condensed or branched thyrses, 1.5–8 cm. long and 0.8–2 cm. wide, the uppermost often clustered to form a lobed panicle, lateral clusters of flowers (condensed or unequally branched cymes) approximate to slightly separated, dense or laxer, up to ± 1.5 cm. long; peduncle up to ± 9 cm., but mostly shorter. Inflorescence-axis often concealed amid the dense flowers, glabrous to variably multicellular hairy. Bracts and bracteoles broadly deltoid-ovate, 2–3 mm., scarious, 1-nerved, glabrous or ± ciliate along the margins and dorsal surface of the nerve with multicellular hairs. Perianth-segments elliptic-oblong, (3–)3.5–5 mm., obtuse to subacute, mucronate with the percurrent midrib, pale and membranous, usually with a narrow central dark vitta at least when dry, with 2–4 usually short and sometimes faint lateral nerves (sometimes reaching halfway or more) on each side of the midrib at the base. Free portion of filaments longer than the sheath, sinuses rounded with no intermediate teeth. Stigmas (2–)3(–4), reflexed, a little shorter than the ± 1 mm. style; ovary multiovulate (mostly ± 13–25). Capsule globose, 2–4 mm., included. Seeds black, shining, compressed, ± 1 mm., deeply furrowed and punctate, the concentric interrupted grooves more conspicuous towards the margins, the punctae towards the middle.

UGANDA. Kigezi District: Malamagambo Forest, Feb. 1950, *Purseglove* 3280!; Mengo District: Bukasa sandpits, Aug. 1952, *Lind* 96! & Entebbe, Aug. 1929, *Maitland* 3863!
TANZANIA. Bukoba District: Kiamawa, Sept./Oct. 1935, *Gillman* 412!; Mwanza District: "presumably on or near Ukerewe Island", *Conrads* 5226!
DISTR. U2, 4; T1; throughout tropical western Africa from Senegal to Angola, also in Sudan, Zaire, Rwanda and Zambia
HAB. In or at the edges of forests; 1060–1520 m.

SYN. [*C. laxa* sensu Gilg in P.O.A. C: 172 (1895); F.T.A. 6(1): 18 (1909); F.P.S. 1: 117 (1950); Hauman in F.C.B. 2: 20 (1951); Cavaco in Mém. Mus. Nat. Hist. Nat. Paris, sér. B: 43 (1962); F.P.U.: 102 (1962) et auctt. mult., *non* Schumach. & Thonn.]

9. **C. hastata** *Lopr.* in Malpighia 14: 427 (1901) & in E.J. 30: 106 (1901); F.T.A. 6(1): 24 (1909); F.D.O.-A. 2: 208 (1932); C.C. Townsend in Hook., Ic. Pl. 38(2): 67, t. 3738 (1975). Type: Tanzania, Tanga District, Moa [Muoa], *Holst* 3124 (COI, K, iso.!)

Erect or scrambling and straggling herb, ± 1–2.5 m. Stem and branches ridged or subterete and striate, moderately to densely furnished with pale or brownish multicellular hairs. Leaves varying from deltoid-ovate with or without broad blunt lateral lobes at the base to hastate with a narrowly oblong central lobe and broad or narrower basal lobes, acute to acuminate, sparsely furnished along the veings of both surfaces (but particularly the lower) with multicellular hairs; lamina of main stem leaves 2–5 × 1.5–3.8 cm., cuneate to subcordate at the base, ± decurrent into the slender, 0.8–2.5 cm. petiole; upper and branch leaves smaller, usually narrower, more shortly stalked. Inflorescences terminal on the stem and branches, of elongate thyrses up to ± 15 × 3 cm., formed of approximated few-many-flowered dense cymose clusters up to ± 1.5 cm. in diameter. Bracts and bracteoles lanceolate-ovate, 4 mm., navicular, firm, straw-coloured or pale brownish with a single darker nerve, minutely denticulate above, often ± ciliate with multicellular hairs below. Perianth-segments narrowly oblong-elliptic, 5–6 mm., ciliate below, pinkish to white or green, in the dry state ultimately straw-coloured to brown, 5–7-nerved almost to the apex along the darker central portion. Free portion of the filaments only ± ¼ the length of the sheath, sinuses without or rarely with very short bluntly deltoid teeth. Stigmas 3, reflexed, slightly shorter than the ± 0.75 mm. style; ovules numerous (15–20 or more). Capsule ovoid, ± 2 mm., included. Seeds ± 1.2 mm., black, shining, compressed, with punctate furrows, the furrows narrower, more crowded and regularly concentric towards the margins.

KENYA. Kilifi District: Kibarani, Mar. 1946, *Jeffery* K. 493! & Malindi, May 1960, *Rawlins* 916! & Sabaki, 6 km. N. of Malindi, Nov. 1961, *Polhill & Paulo* 716!
TANZANIA. Tanga District: Moa [Muoa], July 1893, *Holst* 3124! & Siga Caves, Aug. 1932, *Geilinger* 1293!; Rufiji District: Mchungu Forest Reserve, Oct. 1977, *Wingfield* 4305!
DISTR. **K**7; **T**3, 6; not known elsewhere
HAB. On sandy soil by riverside in coastal bushland and by fringes of forest and *Brachystegia* woodland; 3–106 m.

SYN. [*C. pandurata* sensu Bak. & C.B. Cl. in F.T.A. 6(1): 21 (1909), pro parte *non* Bak.]
C. pandurata Bak. forma *trigyna* Suesseng. in Mitt. Bot. Staats., München 1: 74 (1951). Type: Kenya, Kilifi District, Kibarani, *Jeffery* K. 95 (K, holo.!)

NOTE. One of the above-cited specimens (*Rawlins* 916) alone shows the minute interstaminal teeth, but shows no other differences from all the other specimens seen of this species.
It is extraordinary that in F.T.A., Baker & Clarke gave *C. hastata* as an imperfectly known species, when the Kew isotype of this species was cited, incorrectly identified as *C. pandurata* Baker.

10. **C. elegantissima** *Hauman* in B.J.B.B. 18: 105 (1946) & in F.C.B. 2: 22 (1951); C.C. Townsend in Hook., Ic. Pl. 38 (2) 71, t. 3739 (1975). Type: Zaire, W. of L. Kivu, *Humbert* 7491 (BR, holo.!, K, iso.!)

Erect or scrambling herb with rather few to many long slender branches and with a horizontal, branched rootstock, from 0.6–2 m. and probably more in height. Stem and branches wiry, ridged, glabrous or thinly furnished with short multicellular hairs. Leaves ovate or broadly to narrowly lanceolate-ovate, ± long-acuminate; lamina of the main stem-leaves 5–13 × 2.25–5 cm., shortly cuneate to subtruncate at the base and tapering into the ± 1.5–2.5 cm. petiole; upper and branch leaves smaller, somewhat narrower, more shortly petiolate. Inflorescences axillary and terminal, of simple or branched elongate thyrses, the upper forming a broad spreading panicle up to ± 25 × 26 cm., the slender axes subglabrous or thinly furnished with yellowish multicellular hairs; lateral cymose clusters of flowers few-flowered, lax with branches up to ± 2 mm. long, to rather dense and subsessile, rather distant to subapproximate, ± 4–8 mm. in diameter. Bracts and bracteoles deltoid-ovate, acuminate, 2–2.5 mm., hyaline with a pale or brownish midrib, glabrous or ± furnished with multicellular hairs. Perianth-segments oblong-elliptic, 2–3 mm., silvery-hyaline with a slightly denser, white, narrow central vitta along the single minutely percurrent midrib, and occasionally with 2–5 very faint, short lateral nerves at the base. Free portion of the (? always) magenta filaments subequalling the sheath, with no intermediate teeth. Stigmas 2–3, recurved, subequalling the ± 0.75 mm. style; ovary (3–)5–6(–8)-

ovulate. Capsule ovoid, included or slightly exserted, 2–2.25 mm. Seeds ± 1 mm., black, compressed, shining, furnished with punctate grooves which become shorter and narrower and ± concentric towards the outer edge.

UGANDA. Kigezi District: Impenetrable Forest, May 1939, *Purseglove* 732!
TANZANIA. Kigoma District: Kabogo Head, June 1963, *Kyoto Univ. Exped.* 399!; Mpanda District: Pasagulu, Aug. 1959, *Harley* 9187! & Kasangazi, July 1958, *Mgaza* in *Jefford, Juniper & Newbould* 176!
DISTR. U2; T4; Zaire, Burundi, Rwanda
HAB. In forest and by rivers; 830–1820 m.

11. **C. globosa** *Schinz* in Viert. Nat. Ges. Zürich 76: 137 (1931); Hauman in F.C.B. 2: 23 (1951); Cavaco in Mém. Mus. Nat. Hist. Nat. Paris, sér. B, 13: 46 (1962); C.C. Townsend in Hook., Ic. Pl. 38(2): 75, t. 3740 (1975). Type: Zaire, Likimi, *Goossens* 4719 (BR, holo.!)

Annual herb, erect or straggling, 10–70 cm. Stem and branches ridged or sulcate-striate, glabrous or sparingly furnished with fine multicellular hairs. Leaves broadly elliptic to elliptic-ovate, subacute to rather long-acuminate, glabrous or with scattered fine multicellular hairs on the lower surface; lamina of main stem leaves 3.5–15×2–7 cm., cuneate to attenuate at the base, ± decurrent into a 0.4–2.5 cm. petiole; upper and branch leaves smaller, often narrower. Inflorescences silvery-white to straw-coloured or reddish, terminal and axillary, tightly globose and ± 1–2(–3) cm. in diameter, or laxer but roundish in outline, or rarely the terminal spike-like; all sessile, or shortly pedunculate with peduncles rarely exceeding 1.5 cm. in length. Bracts and bracteoles lanceolate, acute, 2–3 mm., glabrous or ciliate, mucronate with the percurrent nerve. Perianth-segments lanceolate-oblong, 3–4 mm., ± tapering to a ± acute apex, mucronate with the percurrent midrib, which may be solitary or flanked at the base with 1–2 short faint lateral nerves on either side. Free portion of filaments longer than the sheath, sinuses with no intermediate teeth. Stigmas (2–)3, slender, slightly exceeding to twice as long as the ± 0.5–0.7 mm. style; ovary 6–11-ovulate. Capsule globose to shortly ovoid, ± 3 mm., included. Seeds similar to those of *C. isertii*.

var. **globosa**

Inflorescence dense, compact, globose, sessile or shortly pedunculate, or laxer but still ± rounded in outline. Inflorescence silvery-white to reddish.

UGANDA. Masaka District: Towa Forest Reserve, July 1945, *Purseglove* 1747!; Mengo District: Entebbe, Sept. 1922, *Maitland* 204! & Damba I., Nov. 1949, *Dawkins* 458!
DISTR. U4; Zaire, Cameroun, Nigeria

SYN. *C. pandurata* Bak. var. *elobata* Suesseng. in Mitt. Bot. Staats., München 1: 104 (1952). Type: Uganda, Mengo District, near Entebbe, *Dawkins* 544 (K, holo.!)

var. **porphyrostachya** *C.C. Townsend* in Hook., Ic. Pl. 38(2); 76, t. 3740 (1975). Type: Uganda, Mengo District, Kyabana, *Dummer* 2596 [K, holo.!)

Terminal inflorescence spiciform, up to ± 3 × 1 cm. Inflorescences all brown to reddish.

UGANDA. Ankole District: 145 km. on Masha road, May 1937, *Chandler* 1627!; Mengo District: Mukono, Oct. 1931, *Hansford* 2288! & Mpanga Forest, 15 Nov. 1957, *Lind* 2227!
DISTR. U2, 4; Zaire
HAB. (of species as a whole). In forest shade along paths and in woodland dells, by clearings, or in or by forest swamps, sometimes along roadsides, apparently on poor soil with little ground cover; 1150–1210 m.

NOTE. According to Dawkins the species is common throughout forests in the Mengo District but rarely flowers profusely.

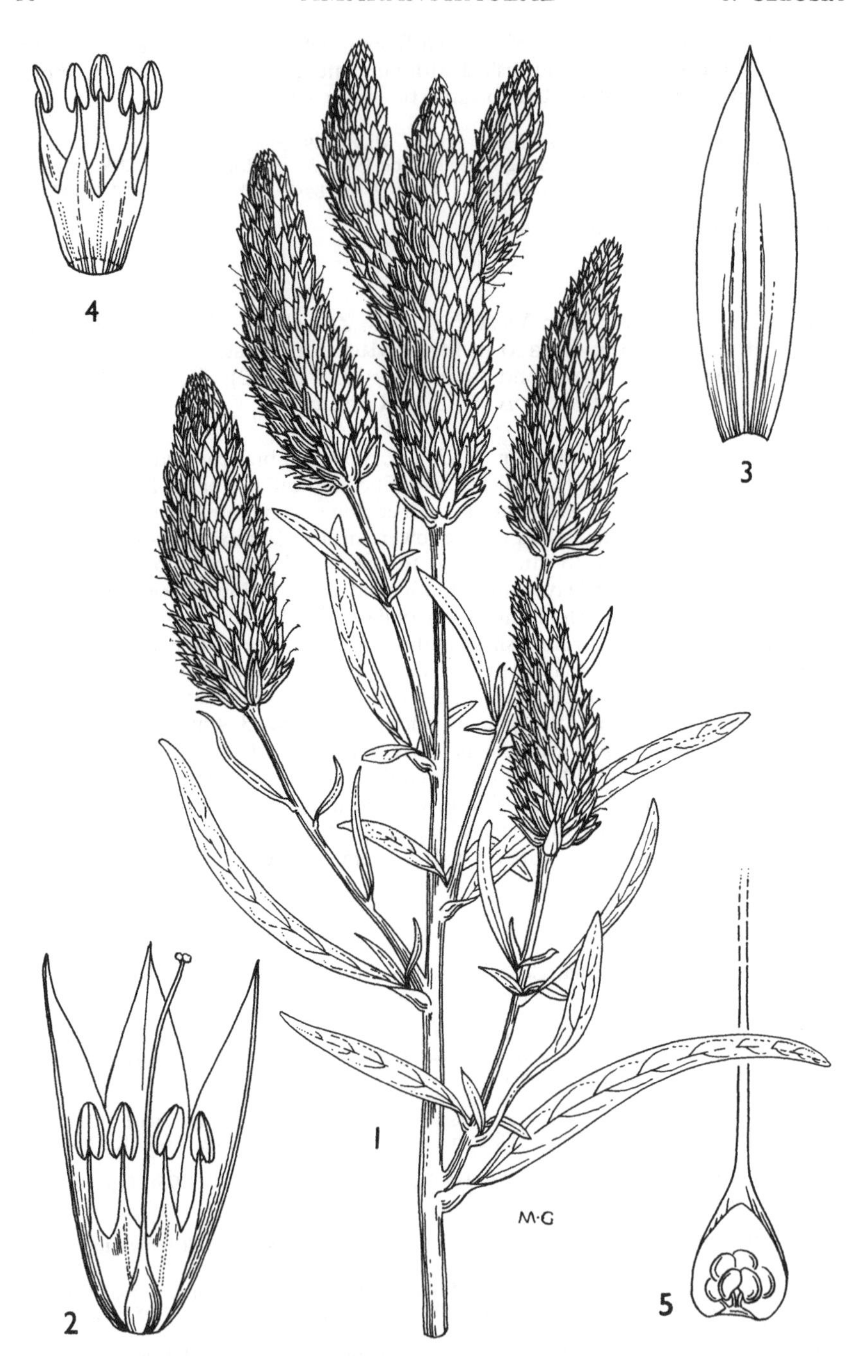

FIG. 2. *CELOSIA ARGENTEA* —**1**, flowering branch, × $\frac{2}{3}$; **2**, flower with two tepals and part of androecium removed, ×6; **3**, tepal, × 6; **4**, androecium, × 6; **5**, gynoecium, longitudinal section, × 6. All from *Chandler* 164. Drawn by Mary Grierson; reproduced from 'Hooker's Icones Plantarum'.

12. **C. argentea** *L.*, Sp. Pl.: 205 (1753); P.O.A. C: 172 (1895); F.T.A. 6(1): 17 (1909); F.D.O.-A. 2: 208 (1932); F.P.S. 1: 117 (1950); Hauman in F.C.B. 2: 16 (1951); E.P.A.: 55 (1953); Cavaco in Mém. Mus. Nat. Hist. Nat. Paris, sér. B, 13: 41 (1962); U.K. W.F.: 131 (1974); C.C. Townsend in Hook., Ic. Pl. 38(2): 115, t. 3750 (1975). Type: *Herb. Linnaeus* 288.1 (LINN, lecto.!)

Annual herb, erect, 0.4–2 m., simple or with many ascending branches. Stem and branches strongly ridged and often sulcate, quite glabrous. Leaves lanceolate-oblong to narrowly linear, acute to obtuse, shortly mucronate with the excurrent midrib, glabrous; lamina of the leaves from the centre of the main stem 2–15 × 0.1–3.2 cm., tapering below into an indistinctly demarcated, slender petiole; upper and branch leaves smaller, markedly reducing; axils often with small-leaved sterile shoots. Inflorescence a dense (rarely laxer below) many-flowered spike 2.5–20 × 1.5–2.2 cm., silvery to pink, at first conical but becoming cylindrical in full flower, terminal on the stem and branches, on a long sulcate peduncle up to ± 20 cm. long, which often lengthens during flowering. Bracts and bracteoles lanceolate or the lower deltoid, 3–5 mm., hyaline, ± aristate with the excurrent single nerve, persistent after the fall of the flower, as are the 2 similar bracteoles. Perianth-segments 6–10 mm., narrowly elliptic-oblong, acute to rather blunt, shortly mucronate with the excurrent midrib, with 2–4 lateral nerves ascending more than halfway up the centre of each segment, margins widely hyaline. Filaments very delicate, free part subequalling or exceeding the sheath, sinuses with no or very minute intermediate teeth; anthers and filaments creamy to magenta. Stigmas 2–3, very short, the filiform style 5–7 mm. long; ovary 4–8-ovulate. Capsule ovoid to almost globular, 3–4 mm. Seeds lenticular, ± 1.25–1.5 mm., black, shining, with a fine reticulate pattern. Fig. 2.

UGANDA. W. Nile District: Pakwach, July 1940, *Purseglove* 971!; Teso District: Serere, *Chandler* 164!; Mengo District: Mukono, Sept. 1913, *Dummer* 226!

KENYA. Northern Frontier Province: Mbalambala [Balambala], Jan. 1947, *J. Adamson* 355! Masai District: Uaso Nyiro [Ngare nyiro], Sept. 1944, *Bally* 3808!; Tana River District: Mlango ya Simba, Nov. 1957, *Greenway & Rawlins* 9461!

TANZANIA. Mwanza, Apr. 1926, *Davis* 251!; ? Ufipa District: Rukwa, Mar. 1959, *Richards* 12223!; Iringa District: Pawaga, May/June 1936, *Ward* P. 9!; Zanzibar I., Mkadini, 7 Mar. 1931, *Vaughan* 1884!

DISTR. **U**1, 3, 4; **K**1, 2, 4, 6, 7; **T**1–5, 7; **Z**; practically a pantropical weed, possibly originating in tropical Africa, where it is widespread

HAB. Weed of cultivation, or in flood plains or along rivers, ranging to grassland or dry lava hills; 610–1640 m.

SYN. *C. debilis* S. Moore in J.B. 54: 291 (1916). Type: Uganda, Mengo District, Kyabana, *Dummer* 2647 (BM, holo.!)
[*Hermbstaedtia argenteiformis* sensu Peter in F.D.O.-A. 2: 210 (1932), *non* Schinz]

NOTE. An easily recognised species, varying little except in leaf shape and (more rarely) in density of the inflorescence.

The forma *cristata* (L.) Schinz is the "cockscomb" form in which the upper part or almost the whole of the inflorescence produces sterile shoots with narrower bracts and bracteoles. It varies much in form, the sterile shoots being long and plumose to short and stumpy, and in many colour forms it is widely cultivated in the tropics – including no doubt the area of the Flora, though no specimens have been seen. An account of the probable origin and relationships of this plant is given by Khoshoo & Pal in J.L.S. 66: 127–141 (1973).

2. HERMBSTAEDTIA

Reichenb., Consp. Regn. Veg.: 164 (1828); C.C. Townsend in K.B. 37: 81–90 (1982)

Berzelia Mart. in Nov. Act. Acad. Caes.-Leop. Carol., Nat. Curios. 13(1): 292 (1826), *non* Brongn. (1826)
Hyparete Raf., Fl. Tellur. 3: 43 (1837)
Langia Endl., Gen. Pl: 304 (1837)

Annual or perennial herbs, sometimes woody at the base, but never scandent. Leaves alternate, simple, entire. Flowers small, bibracteolate, in dense to elongating bracteate spikes or occasionally rounded capitula, ⚥, usually white or pink. Perianth-segments 5, free, equal. Stamens 5, fused into a tube below, the short antheriferous teeth alternating with conspicuous pseudostaminodes, with a long or short tooth on each side, or set on the rounded apex of the dilated filaments; anthers bilocular. Ovary with numerous ovules; style short and ill-marked or occasionally distinct; stigmas (2–)3(–5). Capsule ovoid, circumscissile, never thickened above. Seeds black, compressed-lenticular, ±shining, reticulate.

14 species, chiefly in Namibia and South Africa, with only one outlying species in East Africa.

1. **H. gregoryi** *C.B. Cl.* in F.T.A. 6(1): 26 (1909). Type: Kenya, Tana River District, Tana R. plains, Lake Dumi; *Gregory* (BM, holo.!)

Procumbent (? perennial) herb with long rather straggling stems up to ± 30 cm. long, irregularly branched. Stems and branches wiry, striate and slightly angular, glabrous. Leaves spathulate-obovate, 8–20 × 6–15 mm., thick and rather succulent, with only 2–4 obscure lateral nerves, obtuse to slightly emarginate at the apex, narrowed but not petiolate at the base. Inflorescence very dense, subglobose to shortly cylindrical, 1–4.5 × 1–1.25 cm., white, set close above the congested uppermost leaves. Bracts ovate, 2–2.5 mm., slightly mucronate with the excurrent midrib; bracteoles similar or more deltoid-ovate in outline. Perianth-segments elliptic-oblong, 4–5 mm., 5-nerved with 1 or more of the lateral nerves frequently branched below, lateral nerves ceasing below the apex, which is obtuse and minutely mucronate with the shortly excurrent midrib. Androecium 4 mm. long, the pseudo-staminodes bifid almost to the base with broadly subulate lobes and slightly overtopping the anthers; anther connective central and district. Stigmas 3; ovary ellipsoid, tapering above into the indistinct style, 6-ovulate. Seeds ± 1 mm., compressed, black, strongly reticulate. Fig. 3.

KENYA. Kilifi District: 1–2 km. N. of Kilifi, Ungwana [Formosa] Bay, 5 Sept. 1974, *Frazier* 1101! & N. Malindi beach at mouth of the Sabaki R., 2 Mar. 1959, *Moomaw* 1562!; Kilifi/Lamu District: N. of Mombasa to Lamu and Witu, 1902, *Whyte*!
DISTR. **K**7; not known elsewhere
HAB. The only habitats recorded for this apparently rare endemic are 'face of a coastal sand-dune that rises to 20 m., in a slight lee but still wind-blown', and 'behind mangroves'

3. AMARANTHUS

L., Sp. Pl.: 989 (1753) & Gen. Pl., ed. 5: 427 (1754)

Annual or more rarely perennial herbs, glabrous or furnished with short and gland-like or multicellular hairs, dioecious (not in E. Africa) or monoecious. Leaves alternate, long-petiolate, simple and entire or sinuate. Inflorescence basically cymose, bracteate, consisting entirely of dense to lax axillary clusters or the upper clusters leafless and ± approximate to form a lax or dense "spike" or panicle. Flowers bibracteolate; perianth-segments (2–)3–5, free or connate at the base, membranous, those of ♀ slightly accrescent in fruit. Stamens free, usually similar in number to the perianth-segments; anthers bilocular. Stigmas 2–3; ovule solitary, erect. Fruit a dry capsule, indehiscent, irregularly rupturing or commonly dehiscing by a circumscissile lid. Seeds usually black and shining; testa thin; embryo annular, endosperm present.

About 60 species in the tropical and warmer temperate regions of both hemispheres, impermanent and casual in cooler temperate regions.

A difficult genus in which matters have been complicated by the ancient cultivation of numerous species (e.g. *A. tricolor*, *A. lividus*) as spinach, and others (e.g. *A. hybridus*, *A.*

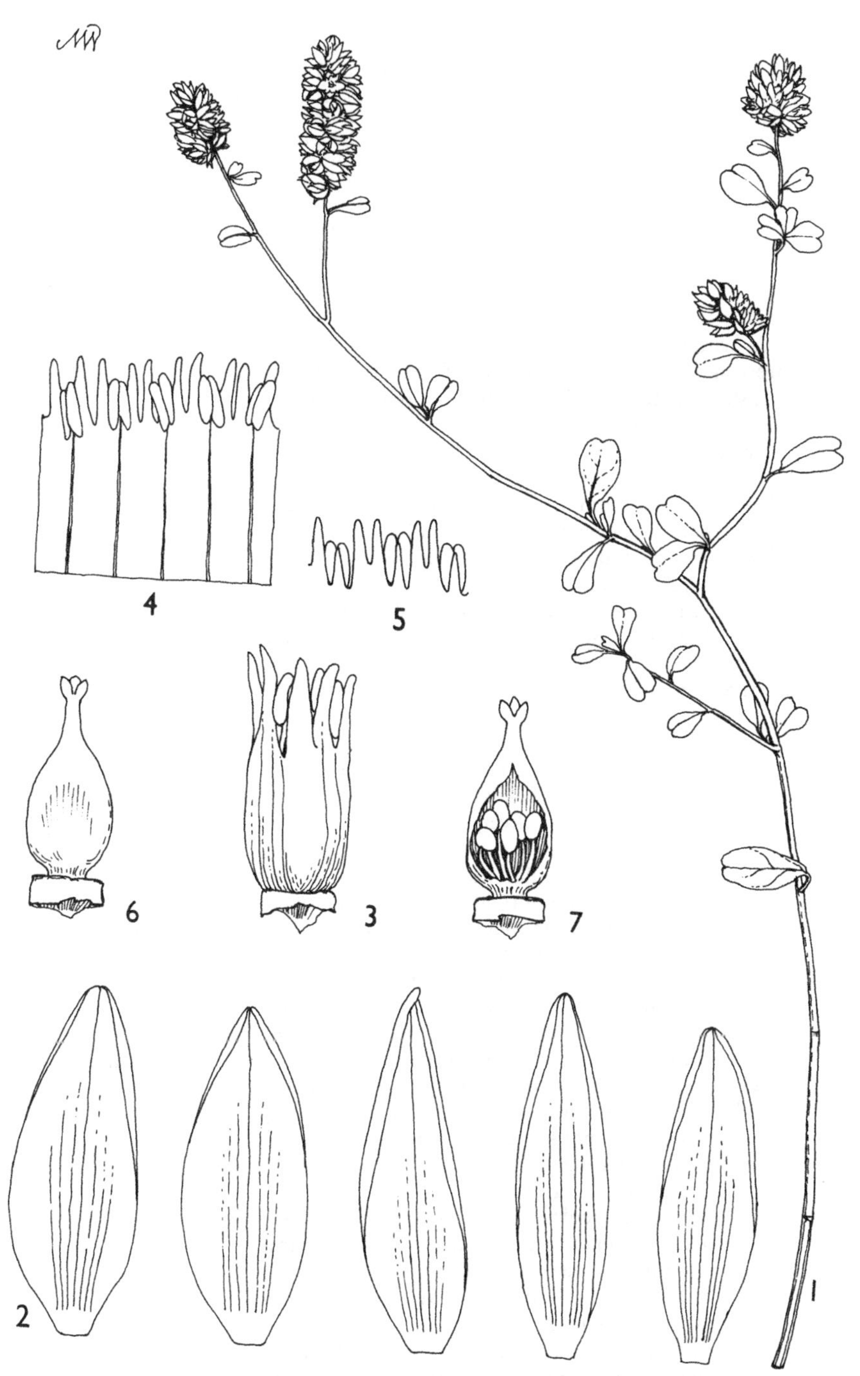

FIG. 3. *HERMBSTAEDTIA GREGORYI*—**1**, flowering stem, × $\frac{2}{3}$; **2**, tepals (outer on left), × 10; **3**, androecium, × 14; **4**, staminal tube opened up, outer surface, × 14; **5**, same apex, inner surface, × 14. **6**, gynoecium, × 14; **7**, ovary opened to show ovules, × 14; All from *Frazier* 1101. Drawn by Mary Millar Watt.

caudatus) as grain crops. The form of the inflorescence, the number and shape of the perianth-segments (particularly of the female flowers) and their length relative to the bracteoles, are all important characters. The pattern on the seed testa is very faint, but systematic examination may well prove that it will supply as useful characters as in *Chenopodium*.

The classic account of the genus is that of Thellung in Ascherson & Graebner, Synopsis der Mitteleuropaischen Flora 5: 225–356 (1914); even if a high proportion of the infraspecific taxa are scarcely worth formal recognition, it remains an invaluable work. Aellen gives a more modern treatment in the second edition of Hegi's Illustrierte Flora von Mittleuropa (Bd. 3, 2, hfg. 1: 465–516, 1959); and Sauer's "The Grain Amaranths and their relatives; a revised taxonomic and geographic survey", in Ann. Missouri Bot. Gard. 54: 103–137 (1967), though accepting a narrower species concept than the present writer is prepared to allow, is also a very useful contribution to our knowledge of the group.

Leaf-axils with paired spines .... 3. *A. spinosus*
Leaf-axils without paired spines:
Inflorescence consisting entirely of axillary cymose clusters, no terminal leafless spike or panicle present:
Tepals of ♀ flowers with slender, divergent, usually colourless awns; bracts and bracteoles long-aristate:
Leaves of the main stem broadest near the base; ♀ tepals broadest above the middle .... 5. *A. tricolor*
Leaves of the main stem broadest about the middle; ♀ tepals broadest below the middle .... 6. *A. thunbergii*
Tepals of ♀ flowers obtuse or shortly mucronate; bracts and bracteoles mucronate or with a rigid coloured awn:
Leaves broadly and conspicuously emarginate; capsule distinctly flattened, indehiscent ...... 10. *A. lividus*
Leaves not or narrowly and indistinctly emarginate; capsule not flattened, dehiscent or indehiscent:
Flowers in globular heads; capsules very acute, rigidly stellately divergent; ♀ tepals obtuse (Fig. 6) .... 8. *A. sparganiocephalus*
Flowers not in globular heads; capsules neither acutely beaked not rigidly stellately divergent; ♀ tepals acute or mucronate .... 7. *A. graecizans*
Inflorescence not entirely of axillary cymose clusters, a terminal leafless spike or panicle present:
Tepals of ♀ flowers very broadly obovate or spathulate, distinctly imbricate; terminal spike of inflorescence long and pendulous, forming a "tail" ... 1. *A. caudatus*
Tepals of ♀ flowers narrowly oblong to lanceolate-oblong or narrowly obovate, not or scarcely imbricate; terminal spike of inflorescence erect or somewhat nodding:
Capsule opening by a circumscissile lid:
Bracts and bracteoles with a fine, flexuose, ± hair-like terminal awn; axillary inflorescences globular, or if axillary spikes present then these with globular clusters at the junction with the stem .... 5. *A. tricolor*
Bracts and bracteoles with a stout, rigid terminal awn, sometimes very short; axillary inflorescences spiciform, with no globular cluster at their junction with the stem:
Male flowers confined to a normally quite short length of the tip of each spike, rarely mixed

with ♀ flowers; lid of capsule strongly wrinkled near the line of dehiscence (Fig. 4/11) .... 4. *A. dubius*

Male flowers mixed with ♀ flowers; lid of capsule almost smooth, or at most longitudinally grooved at the base (Fig. 4/7, 9) .... 2. *A. hybridus*

Capsule not opening by a circumscissile lid:

Capsule (fig. 4/8) ± globular, extremely muricate, not or scarcely exceeding the perianth-segments; seeds (Fig. 5/5, 6) with shallow, scurfy verrucae on the reticulate pattern of the testa .... 11. *A. viridis*

Capsule ellipsoid or distinctly compressed, not muricate, distinctly exceeding the perianth segments; seeds with no shallow verrucae:

Fruit (Fig. 4/1) ellipsoid, scarcely compressed, seed also ellipsoid; leaves rarely (feebly) retuse ... 9. *A. deflexus*

Fruit (Fig. 4/2) compressed, lenticular or shortly pyriform; seeds lenticular; leaves broadly and distinctly emarginate, rarely broadly truncate 10. *A. lividus*

1. **A. caudatus** *L.*, Sp. Pl.: 990 (1753); P.O.A.C: 172 (1895); F.T.A. 6(1): 31 (1909), pro parte; F.D.O.-A. 2: 212 (1932), pro parte; F.P.S. 1: 116 (1950); Hauman in F.C.B. 2: 29 (1951); E.P.A.: 58 (1953); U.K.W.F.: 133 (1974). Type: *Herb. Hort. Cliff.* p. 443 *Amaranthus* 1 (BM, lecto.!)

Annual herb, erect, up to ± 1.5 m., commonly reddish or purplish throughout. Stem rather stout, not or sparingly branched, glabrous or thinly furnished with rather long multicellular hairs which are increasingly numerous upwards. Leaves glabrous, or ± sparingly pilose along the margins and lower surface of the primary venation, long-petiolate (petiole to ± 8 cm. but not longer than the lamina); lamina broadly ovate to rhomboid-ovate or ovate-elliptic, 2.5–15 × 1–8 cm., obtuse to subacute at the mucronulate tip, shortly cuneate to attenuate below. Flowers in axillary and terminal spikes formed of increasingly approximated cymose clusters, the terminal inflorescence varying from a single, elongate tail-like pendulous spike, to 30 cm. or more long and ± 1.5 cm. wide, to a panicle with the ultimate spike so formed; ♂ and ♀ flowers intermixed throughout the spikes. Bracts and bracteoles deltoid-ovate, pale and membranous, acuminate and with a long pale or reddish rigid erect arista formed by the yellow-green or reddish stout excurrent midrib, the longest bracteoles up to twice as long as the perianth. Perianth-segments 5; ♂ oblong-elliptic, 2.5–3.5 mm., acute, aristate; ♀ broadly obovate to spathulate, 1.75–2.5 mm., distinctly imbricate, abruptly narrowed to a blunt or sometimes faintly emarginate mucronate tip. Stigmas 3, ± 0.75 mm., erect or flexuose. Capsule ovoid-globose, 2–2.5 mm., circumscissile, slightly urceolate, the lid smooth or furrowed below, abruptly narrowed to a short thick beak. Seeds shining, compressed, black, almost smooth and shining, or commonly subspherical, with a thick yellowish margin and a translucent centre, ± 0.75–1.25 mm. Fig. 4/4.

UGANDA. W. Nile District: Paidha, Mar. 1934, *Tothill* 2571!; Ankole District: Kamatalisi, Nov. 1950, *Jarrett* 127!; Busoga District: Mutai, July 1945, *Kibuka* J4!

KENYA. Nairobi, Aug. 1934, *Goldschmidt!*

DISTR. **U** 1–3; **K** 4, throughout much of the tropics and impermanently in some temperate regions, apparently always as an escape from or relic of cultivation as a grain crop or ornamental

HAB. Weed (or relic) of cultivation, valley grassland; 1360–1700 m.

NOTE Unknown in the truly wild state, it has been suggested that this species may have originated in cultivation as a derivative of *A. quitensis* Kunth.

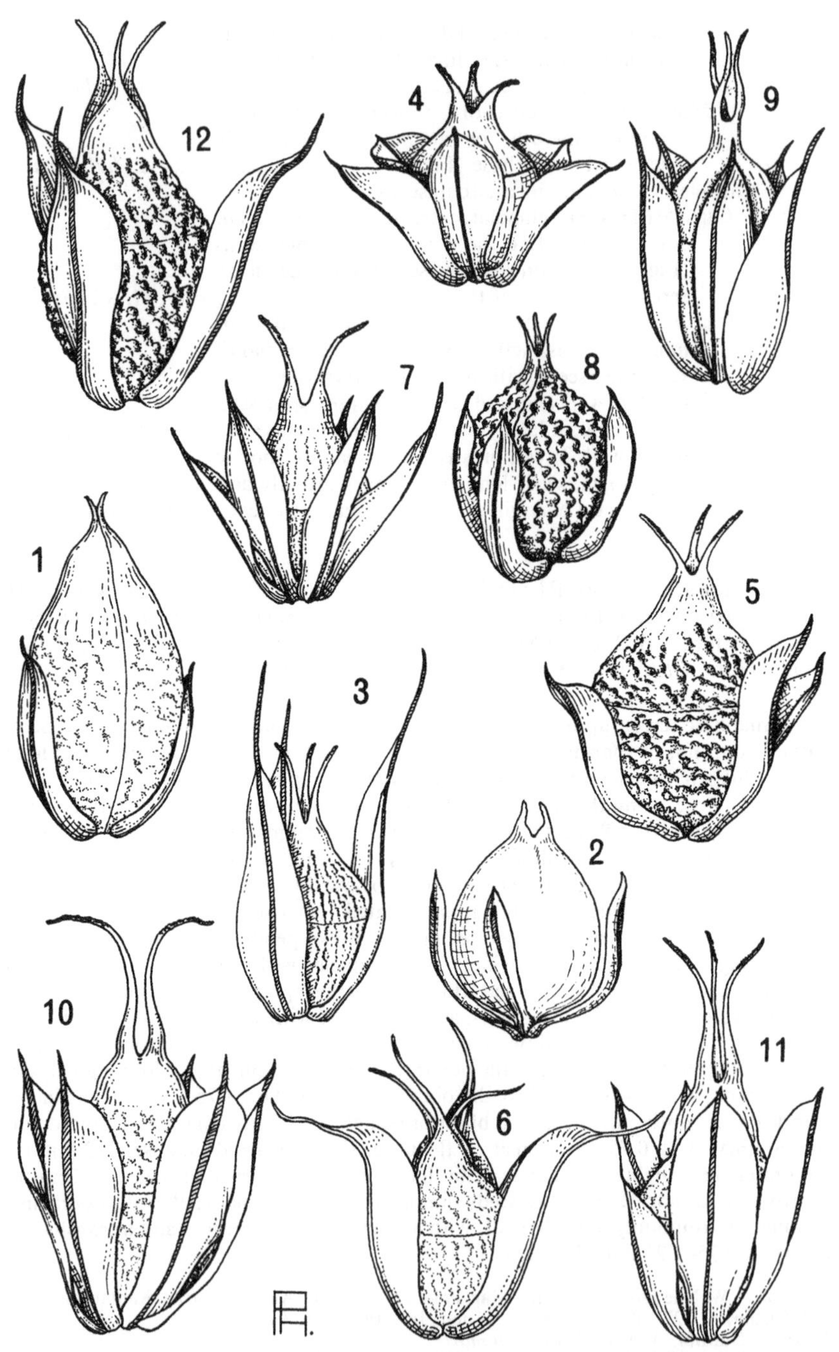

FIG. 4. Fruits of *AMARANTHUS*—**1,** *A. deflexus,* × 20 (*Bally* 8067); **2,** *A. lividus* subsp. *polygonoides,* × 20 (*Archbold* 643); **3,** *A. thunbergii,* × 10 (*J. Wilson* 621); **4,** *A. caudatus,* × 10 (*Tothill* 2571); **5,** *A. graecizans* subsp. *sylvestris,* × 20 (*Dummer* 471); **6,** *A. tricolor,* × 6 (*Greenway* 980); **7,** *A. hybridus* subsp. *hybridus,* × 10 (*Lugard* 638); **8,** *A. viridis,* × 20 (*Richards* 10900); **9,** *A. hybridus* subsp. *cruentus,* × 20 (*Culwick* 4); **10,** *A. spinosus,* × 20 (*Bally* 7828); **11**; *A. dubius,* × 20 (*Conrads* 13288); **12,** *A. graecizans* subsp. *thellungianus,* × 20 (*Bally* 914). Drawn by Pat Halliday.

2. **A. hybridus** *L.*, Sp. Pl.: 990 (1753); F.P.S. 1: 116 (1950); U.K.W.F.: 133 (1974). Type *Herb. Linnaeus* 1117.19 (LINN, lecto.!)

Annual herb, erect or less commonly ascending, up to ± 2(–3) m. in cultivated forms, but much less in spontaneous plants, not infrequently reddish tinted throughout. Stems stout, branched, angular, glabrous or thinly to moderately furnished with short or long multicellular hairs (increasingly so above, especially in the inflorescence). Leaves glabrous, or thinly pilose on the lower margins and underside of the primary nervation, long-petiolate (petioles up to 15 cm. but even then scarcely exceeding the lamina); lamina broadly lanceolate to rhomboid or ovate, 3–19(–30) × 1.5–8(–12) cm., gradually narrowed to the blunt to subacute mucronulate tip, attenuate or shortly cuneate into the petiole below. Flowers in yellowish, green, reddish or purple axillary and terminal spikes formed of cymose clusters, which are increasingly closely approximate upwards, the terminal inflorescence varying from a single spike to a broad much-branched panicle up to ± 45 × 25 cm., the ultimate spike not infrequently nodding; ♂ and ♀ flowers intermixed throughout the spikes. Bracts and bracteoles deltoid-ovate to deltoid-lanceolate, pale and membranous, acuminate and with a long pale to reddish-tipped erect arista formed by the stout excurrent yellow or greenish midrib; bracteoles subequalling to much exceeding the perianth. Perianth-segments 5, lanceolate or oblong, 1.5–3.5 mm., acute-aristate or the inner in ♀ sometimes blunt, only the midrib at most greenish. Stigmas (2–)3, erect, flexuose or recurved, ± 0.75–1.25 mm. Capsule subglobose to ovoid to ovoid-urceolate, 2–3 mm., circumscissile, with a moderately distinct to obsolete beak, lid smooth, longitudinally sulcate or sometimes rugulose below the neck. Seed black and shining, or pale, compressed, 0.75–1.25 mm., almost smooth centrally, faintly reticulate around the margins.

subsp. **hybridus**

Stigma-bases and upper part of lid of fruit swollen, so that the fruit has a distinct inflated beak. Inner perianth-segments of ♀ flowers commonly acute. Longer bracteoles of ♀ flowers mostly about twice as long as the perianth. Fig. 4/7.

UGANDA. Kigezi District: Kachwekano Farm, Dec. 1949, *Purseglove* 3167!; Masaka District: Buwunga (cult.), July 1946, *A.S. Thomas* 4494! & Kakuto (cult.), Aug. 1945, *Purseglove* 1775!
KENYA. Naivasha District: shore of Lake Naivasha, July 1952, *Bally* 8240!; Meru suburbs, May 1951, *Hancock* 39!; Kericho District: Sambret Catchment, Jan. 1961, *Kerfoot* 759!
TANZANIA. Mbulu District: Mbulumbul, June 1945, *Greenway* 7443!; Lushoto District: Mombo–Soni road, June 1953, *Drummond & Hemsley* 3005!; Songea District: Matengo Hills, Mar. 1956, *Milne-Redhead & Taylor* 8789!
DISTR. **U** 2, 4; **K** 1, 3–6; **T** 2, 3, 5, 7, 8; throughout the tropical and subtropical regions of the world, also occurring as a casual in temperate regions, e.g. a common wool adventive in Europe; originated in America
HAB. Commonly as a weed of current or abandoned cultivation or on waste ground, also along roadsides, river margins and forest edges; 900–2210 m.

SYN. *A. hypochondriacus* L., Sp. Pl.: 991 (1753); E.P.A.: 58 (1953). Type: *Herb. Linnaeus* 1117.24 (LINN, lecto.!)
A. chlorostachys Willd. Hist. Amaranth.: 34, t. x/19 (1790). Type: *Herb. Willdenow* 17521 (B, holo., IDC microfiche 1265. 10!)
A. patulus Bertol., Comment. It. Neap. 19, t. 2 (1837). Type: Italy, Naples, 1834 *Bertoloni* (BOLO, holo.!)
A. incurvatus Gren. & Godr., Fl. France Prosp.: 8 (1846). Type: France, Lyon, 1845 *Timeroy* (P, holo.!)
A. powellii S. Wats., Proc. Amer. Acad. 10: 347 (1875). Type: cultivated material grown from Arizona seed collected by Powell (US, holo.)
A. hybridus L. subsp. *hypochondriacus* (L.) Thell. in Fl. Adv. Montpellier: 204 (1912); Hauman in F.C.B. 2: 29 (1951)
A. hybridus L. subsp. *hypochondriacus* (L.) Thell. var *chlorostachys* (Willd.) Thell. in Fl. Adv. Montpellier: 205 (1912)
A. hypochondriacus L. var. *amentaceus* Suesseng. in Mitt. Bot. Staats., München 1: 185 (1953). Type: Kenya, *Hancock* 39 (K, holo.!)
A. hybridus L. subsp. *incurvatus* (Gren. & Godr.) Brenan in Watsonia 4: 268 (1961)

NOTE. Cultivated as a vegetable, noted on one sheet by a British collector to be as palatable as the best English spinach.

subsp. **cruentus** *(L.) Thell.* in Fl. Adv. Montpellier: 205 (1912); F.P.N.A. 1: 129 (1948); Hauman in F.C.B. 2: 27 (1951); Cavaco in Mém. Mus. Nat. Hist. Nat. Paris, sér. B, 13: 55 (1962). Type *Herb. Linnaeus* 1117.25 (LINN, lecto.!)

Longer bracteoles of ♀ flowers mostly 1–1.5 times as long as the perianth. Stigma-bases and upper part of lid of fruit scarcely swollen, fruit with a short, smooth beak. Inner perianth-segments of ♀ flowers commonly obtuse. Fig. 4/9.

UGANDA. Mengo District: Kampala, Oct. 1931, *Small* 2300!, Kipayo, Jan. 1914, *Dummer* 590! & Makerere, June 1972, *Katende* 1706!

KENYA. Nairobi, June 1969, *Bally* 13323!; Embu, May 1932, *M.D. Graham* in *A.D.* 1726!; Kilifi District: Ribe, *Wakefield*!

TANZANIA. Mwanza District: Nyegezi, Apr. 1959, *Tanner* 677!; Tanga District: Kisosora, Sept. 1959, *Semsei* 2913!; Mpwapwa District: Matomondo, Mar. 1945, *van Rensburg* 8!; Zanzibar I., *Toms* 39!

DISTR. **U**4; **K**3–5, 7; **T**1–8; **Z**; presumably of Central American origin, now widespread in tropical and subtropical regions of the world; the red form (which is probably of cultivated derivation) is cultivated as an ornamental in temperate regions also, occasionally escaping there.

HAB Commonly as a weed of cultivation, also on broken waste ground, in short grassland and in shaded places at forest edges; 20–1790 m.

SYN. *A. paniculatus* L., Sp. Pl., ed. 2: 1406 (1753); E.P.A.: 59 (1953). Type: *Herb. Linnaeus* 1117.20 (LINN, lecto.!)
A. cruentus L., Syst. Nat., ed. 10, 2: 1269 (1759)
A. paniculatus L. var. *cruentus* (L.) Moq. in DC., Prodr. 13(2): 257 (1849)
[*A. caudatus* sensu Bak. & C.B. Cl. in F.T.A. 6(1): 31 (1909), pro parte; F.D.O.-A. 2: 212 (1932), pro parte, *non* L.]
A. caudatus L. var. *pseudopaniculatus* Suesseng. (incl. forma *oblongotepalus* Suesseng.) in Mitt. Bot. Staats., München 1: 71 (1951). Type of var.: Tanzania, Lushoto District, Amani, *Greenway* 993 (EA, holo.!); of forma: Tanzania, Lushoto District, Amani, *Greenway* 6155 (K, iso.!)
A. hypochondriacus L. subsp. *cruentus* (L.) Thell. var *subdubius* Suesseng. in Mitt. Bot. Staats., München 1: 73 (1951). Type: Tanzania, Shinyanga, *Koritschoner* 3062 (K, isolecto.!)
[*A. hybridus* L. subsp. *incurvatus* sensu Brenan in Watsonia 4: 268 (1961); U.K.W.F.: 133 (1974), pro parte, *non Amaranthus incurvatus* Gren. & Godr.]

NOTE. The form with red or purple inflorescences seems almost as common in East Africa as that with green or yellowish inflorescences; many specimens, however, lack colour notes. This subspecies is also cultivated as a vegetable.

3. **A. spinosus** *L.*, Sp. Pl.: 991 (1753); P.O.A.C: 172 (1895); F.T.A. 6(1): 32 (1909); F.D.O.-A. 2: 214 (1932); F.P.S. 1: 116 (1950); Hauman in F.C.B. 2: 31 (1951); E.P.A.: 60 (1953); Cavaco in Mém. Mus. Nat. Hist. Nat. Paris, sér. B, 13: 53 (1962); U.K.W.F.: 132 (1974). Type: *Herb. Linnaeus* 1117.27, from Uppsala Botanic Garden (LINN, lecto.!). No specimen exists in Herb. Hort. Cliff. or in Hermann's Herbarium, as might have been expected from Linnaeus' citations.

Annual herb, erect or slightly decumbent, up to ± 1.5 m. Stem stout, sometimes reddish, usually branched, angular, glabrous or increasingly furnished above (especially in the inflorescence) with long multicellular flocculent hairs. Leaves glabrous or thinly pilose on the lower surface of the primary nervation, long-petiolate (petioles up to ± 9 cm., sometimes longer than the lamina); lamina ovate to rhomboid-ovate, elliptic, lanceolate-oblong or lanceolate, ± 1.5–12 × 0.8–6 cm., subacute or more commonly blunt or retuse at the tip, with a distinct fine colourless mucro, cuneate or attenuate at the base; each leaf-axil bearing a pair of fine and slender to stout and compressed spines up to ± 2.5 cm. long. Flowers green, in the lower part of the plant in axillary clusters 6–15 mm. in diameter; towards the ends of the stem and branches the clusters leafless and approximated to form simple or (the terminal at least) usually branched spikes up to ± 15 cm. long and 1 cm. wide; lower clusters entirely ♀ as are the lower flowers of the spikes; upper flowers of the spikes ♂, mostly for the apical ½–⅔ of each spike. Bracts and bracteoles deltoid-ovate, pale and membranous, with an erect reddish awn formed by the excurrent midrib; bracteoles shorter than, subequalling or a little exceeding the perianth, commonly smaller than the bracts. Perianth-segments 5, narrowly oblong or spathulate-oblong, 1.5–2.5 mm., obtuse or acute, mucronulate, frequently with a greenish dorsal vitta; ♂ broadly lanceolate or lanceolate-oblong, acute or acuminate, only the midrib green. Stigmas (2–)3, flexuose or reflexed, 1–1.5 mm. Capsule ovoid-urceolate with a short inflated beak below the style-base ± 1.5 mm., regularly or irregularly circumscissile or rarely indehiscent, the lid rugulose below the neck. Seed 0.75–1 mm., compressed, black, shining, very faintly reticulate. Fig. 4/10.

UGANDA. Karamoja District: Amaler, Jan. 1936, *Eggeling* 2595!; Teso District: Serere, Aug. 1932, *Chandler* 881!; Busoga District: Buteleja Camp, July 1926, *Maitland* 1096!
KENYA. Trans-Nzoia District: Kitale, Aug. 1968, *Tweedie* 3562!; Nairobi, Feb. 1915, *Dummer* 1930!; Masai District: Ngerende, Mar. 1961, *Glover, Gwynne & Samuel* 183!
TANZANIA. Mwanza, Dec. 1964, *Carmichael* 1162!; Lushoto District: Mombo–Soni road, June 1953, *Drummond & Hemsley* 3006!; Uzaramo District: Oyster Bay, June 1968, *Batty* 141!; Zanzibar I., Oct. 1873, *Hildebrandt* 1037!; Pemba I., Chake Chake, Aug. 1929, *Vaughan* 561!
DISTR. **U** 1, 3, 4; **K** 1–7; **T** 1–3, 6–8; **Z** ; **P**; throughout the tropical and subtropical regions of the world, occurring occasionally as a casual in temperate regions also; presumed to be of American origin
HAB. Commonly as a weed of cultivation, waste ground and roadsides, also occurring in open grassland and in wet ground by swamps and along rivers; 0–1820 m.

4. **A. dubius** *Thell.,* Fl. Adv. Montpellier: 203 (1912); * Hauman in F.C.B. 2: 30 (1951). Type – see Townsend in K.B. 29: 471–2 (1974): cultivated material from Erlangen Botanic Garden, ex herb. Schwaegrichen (M, neo.!)

Erect annual herb, up to 0.9(–1.5) m. Stem rather slender to stout, usually branched, angular, glabrous or increasingly furnished upwards (especially in the inflorescence) with short to rather long multicellular hairs. Leaves glabrous or thinly and shortly pilose on the lower surface of the primary venation, long-petiolate (petioles up to ± 8.5 cm., sometimes longer than the lamina); lamina ovate or rhomboid-ovate, 1.5–8 (–12) × 0.7–5 (–8) cm., blunt or retuse at the tip with a distinct fine mucro formed by the percurrent nerve, cuneate (usually shortly so) at the base; leaf-axils without spines. Flowers green, in the lower part of the plant in axillary clusters 4–10 cm. in diameter, towards the ends of the stem and branches the leafless clusters approximated to form simple or (the terminal at least) branched spikes ± 3–15(–25) cm. long and 6–8(–10) mm. wide. Lower clusters of flowers entirely ♀ the spikes generally showing a few ♂ flowers at the tips only (rarely in more than the apical 1 cm.) occasionally with ♂ flowers also scattered among the lower ♀ flowers. Bracts and bracteoles deltoid-ovate, pale and membranous, with an erect reddish awn formed by the excurrent midrib; bracteoles shorter than or subequalling the perianth, rarely slightly exceeding it. Perianth-segments (4–)5; ♀ narrowly oblong or spathulate-oblong, ± 1.5–2.75 mm., obtuse or sometimes (particularly those approaching the ♂ flowers) acute, mucronulate, frequently with a greenish dorsal vitta above; ♂ broadly lanceolate or lanceolate-oblong, generally acuminate, only the thin midrib green. Stigmas 3, flexuose or reflexed, ± 0.75–1 mm. Capsule ovoid-urceolate with a short inflated beak below the style base, ± 1.5–1.75 mm., circumscissile, the lid strongly rugulose below the neck. Seed 1–1.25 mm., compressed, black, shining, faintly reticulate. Fig. 4/11.

UGANDA. Karamoja District: Moroto, May 1971, *J. Wilson* 2082!; Bunyoro, Feb. 1943, *Purseglove* 1204!; Mengo District: Kajansi Forest, May 1935, *Chandler* 1208!
KEYNA. Northern Frontier Province: Moyale, July 1952, *Gillett* 13636!; Turkana District: Lorengipe, Oct. 1963, *Bogdan* 5648!; Nairobi, Feb. 1915, *Dummer* 1928!
TANZANIA Mbulu District: Chem Chem R., June 1965, *Greenway & Kanuri* 11891; Mpanda District: Kibwesa Point, July 1958, *Juniper & Jefford* 64!; Dar es Salaam, Oct. 1967, *Harris*

*The name was originally proposed without description by Martius, Pl. Hort. Acad. Erlang.: 197 (1814).

1080!; Zanzibar I., Mazizini, Oct. 1962, *Faulkner* 3105!; Pemba I., Chake Chake, Aug. 1929, *Vaughan* 562!

DISTR. **U**1–4; **K**1, 2, 4, 7; **T**8; **Z**; **P**; of American origin, now found practically throughout the tropical regions of the world, scarcer as an adventive in temperate regions than *A. spinosus*

HAB. Weed of cultivation, waste places near villages and by roadsides, in short grassland and in stony places on hillsides, cleared areas in forests, silty riversides; 0–1660 m.

SYN. [*A tristis* sensu Moq. in DC., Prodr. 13(2): 260 (1849), *non* L.]
[*A. patulus* sensu Bak. & C.B. Cl. in F.T.A. 6(1): 33 (1909), *non* Bertol.]
A. hybridus L. var. *cruentus* (L.) Thell. forma *acicularis* Suesseng. in Mitt. Bot. Staats., München 1:4 (1950). Type: Uganda, Masaka District, Kabula, *Purseglove* 1813 (K, holo.!)
A. dubius Thell var. *crassespicatus* Suesseng. in Mitt. Bot. Staats., München 1: 73 (1951). Type: Tanzania, Chunya District, Lupa [Mlupa], *Geilinger* 3011 (K, isolecto.!)

NOTE. *A. dubius* is the only known polyploid *Amaranthus*, and it is postulated by Grant, Canadian Journ. Bot. 37: 1063–70 (1959), that it arose as an allotetraploid with *A. spinosus* as one parent and possibly *A. quitensis* as the other – a conclusion disputed by Pal & Khoshoo, Curr. Sci. 34: 370–371 (1965). Hybrids between *A. dubius* and *A. spinosus* appear to occur freely where these two species are associated, and Srivasta, Pal & Nair, Rev. Palaeobot. Palynol. 23: 287–291 (1977), claim that these may be distinguished by the presence of micrograins among the pollen. Such hybrids will certainly occur in tropical Africa. One specimen, *Sanane* 214 from Misumhumilo, Tanzania (T4, Mpanda District) has very reduced spines and only a few male flowers at the tips of the spikes; it could well be one example of a hybrid of this parentage. In spite of various characters used in the literature, I have been able to find no infallible means by which *A. spinosus* can be separated from *A. dubius* other than by the presence or absence of spines, though the generally considerably greater number of terminal male flowers in the spikes of *A. spinosus* seems a reasonably reliable character where only "tops" are collected. According to Srivasta, Pal & Nair (l.c.) the pollen of *A. dubius* has larger pores than that of *A. spinosus*.

5. **A. tricolor** *L.*, Sp. Pl.: 989 (1753); F.T.A. 6(1): 32 (1909); F.D.O.-A. 2: 213 (1932); F.P.S. 1: 116 (1950); E.P.A.: 61 (1953). Type: *Herb. Linnaeus* 1117.7 (LINN, lecto.!)

Annual herb, ascending or erect, attaining ± 1.25 m. in cultivation in Africa (even more in tropical Asia). Stem stout, usually much branched, it and the branches angular, glabrous or furnished in the upper parts with sparse (or denser in the inflorescence) ± crisped hairs. Leaves glabrous, or thinly pilose on lower surface of primary nervation, green or purplish-suffused, very variable in size, long-petiolate (up to ± 8 cm.) lamina broadly ovate, rhomboid-ovate or broadly elliptic to lanceolate-oblong, emarginate to obtuse or acute at the apex, at the base shortly cuneate to attenuate, decurrent along the petiole. Flowers green to crimson, in ± globose clusters ± 4–25 mm. in diameter, all or only the lower axillary and distant, the upper sometimes without subtending leaves and increasingly approximate to form a thick terminal spike of variable length, ♂ and ♀ flowers intermixed. Bracts and bracteoles broadly ovate or deltoid-ovate; bracteoles subequalling or shorter than the perianth, pale and membranous, broadest near the base and gradually narrowed upwards to the green midrib which is excurrent to form a long pale-tipped awn usually at least half as long as the basal portion and not rarely equalling it. Perianth-segments 3, elliptic or oblong-elliptic, 3–5 mm. long, narrowed above, pale and membranous, the green midrib excurrent into a long pale-tipped awn, those of the ♀ flowers slightly accrescent in fruit. Stigmas 3, erect or recurved, ± 2 mm. Capsule ovoid-urceolate with a short beak below the style-base, 2.25–2.75 mm., circumscissile, membranous, obscurely wrinkled. Seed 1–1.5 mm., black or brown, shining, very faintly reticulate. Fig. 4/6.

TANZANIA. Lushoto District: Amani Nursery, *Greenway* 980!, 5955!, 5956!, 6156–8!

DISTR. Cultivated as a vegetable at Amani, and no doubt elsewhere; apparently a native of tropical Asia, this species is a well-known ornamental and vegetable in the tropics

SYN. *A. tristis* L., Sp. Pl.: 989 (1753). Type: a specimen from Uppsala Botanic Garden, probably *Herb. Linnaeus* 1117. 12, "*indica*" (LINN, lecto.!)
A. inamoenus Willd., Hist. Amaranth.: 14, t. VII/14(1790). Type: *Herb: Willdenow* 17504 (B, holo., IDC microfiche 1263. 24!)

A. tricolor L. var. *tristis* (L.) Thell. in Aschers. & Graebn., Syn. Mitt. Eur. Fl. 5(1): 274 (1914)
A. tricolor L. subsp. *tristis* (L.) Aellen in Hegi, Ill. Fl. Mitt-Eur., ed. 2, 3(2): 495 (1959)

NOTE. The names "Spinach" and "Dwarf Chinese" are noted by Greenway on two of the Amani gatherings (Nos. 5955, 5956).

Under the treatment of Aellen, in Hegi, Ill. Fl. Mitt-Eur., ed. 2, 3(2): 494-496 (1959), the East African material would fall under subsp. *tristis* (L.) Aellen, in which the inflorescence forms a leafless terminal spike in addition to ± globose axillary flower clusters in the lower parts of the stem and branches. I have already commented on the unworkability of this treatment in Fl. W. Pakistan No. 71, Amaranthaceae: 14 (1974), and study of populations of *A. tricolor* in the field in Sri Lanka has confirmed my opinion that *A. tricolor* should be treated as a polymorphic species without employing infraspecific taxa. Provision has been made in the key for the possible existence of depauperate forms of the species with no terminal spike being found in East Africa.

6. **A. thunbergii** *Moq.* in DC., Prodr. 13(2): 262 (1849); Hauman in F.C.B. 2: 32 (1951); E.P.A.: 60 (1953). Type: South Africa, *Herb. Thunberg* 22237, sub *A. albus* (UPS, iso., IDC microfiche 93.6!)*

Annual herb, ascending or erect, simple or branched from the base and frequently also above, 15–55 cm. Stem and branches stout, angular, glabrous or thinly hairy below, upwards increasingly furnished with long crisped multicellular rather flocculent hairs. Leaves glabrous or thinly pilose on the lower surface of the primary nervation, sometimes with a dark purple blotch, long-petiolate (petioles up to ± 4 cm., sometimes longer than the lamina); lamina narrowly or broadly elliptic to rhomboid or spathulate, ± (5–)15–45(–60) × (4–)10–30 (–40) mm., blunt or retuse at the apex with the midrib excurrent in a short mucro, at the base cuneate to attenuate, ± decurrent along the petiole. Flowers green, ♂ flowers most frequent at the top of the upper clusters, all in axillary clusters 6–15 mm. in diameter, usually increasingly distant towards the base of the stem and branches, ♂ and ♀ flowers intermixed. Bracts and bracteoles deltoid-lanceolate, bracts subequalling or slightly exceeding the perianth, pale and membranous, often greenish centrally above, the midrib excurrent in a long fine awn often as long as the basal portion, bracteoles shorter (to 2 mm. long), awn colourless and ± reflexed above. Perianth-segments 3, similar in ♂ and ♀, lanceolate to oblong or rarely narrowly spathulate, 3–6 mm., pale and membranous or (especially in ♀) somewhat greenish above, gradually or more abruptly narrowed into the 0.75–1.5 mm. awn formed by the excurrent midrib, the latter green but the awn colourless above; fruiting perianth-segments slightly accrescent, wider than those of the ♂ flowers. Stigmas 3, flexuous or often reflexed, pale, 1.5–2 mm. Capsule pyriform with a short beak, ± 2.5–3.5 mm., circumscissile, membranous, obscurely wrinkled, shorter than the perianth (attaining ± the base of the aristate tips). Seed 1–1.5 mm., black or chestnut, shining, feebly reticulate, Fig. 4/3.

UGANDA. Karamoja District: near Moruangaberu, June 1957, *Dyson-Hudson* 182!, Namalu, Aug. 1964 *Wilson* 1738!. & Labwor Hills, Apr. 1954, *Wilson* 786!.
KENYA. Kiambu District: Ruiru, *James* in *C.M.* 2304!; Nairobi, *Sturrock* 570 B.
TANZANIA. Chunya, 1937, *Raymond* 109!
DISTR. **U** 1; **K** 4; **T** 7; tropical Africa from Ethiopia and Somalia to Zaire and Angola, through to Namibia and South Africa; introduced into Australia and thence to Europe as a frequent casual wool adventive
HAB. Weed of cultivation, grazed grassland, grassland on rocky soil; 1100–1640 m.

SYN. [*A. graecizans* sensu Bak. & C.B. Cl. in F.T.A. 6(1): 34 (1909), *non* L.]
A. sp. A. sensu Agnew, U.K.W.F.: 133 (1974)

NOTE. *Dyson-Hudson* 182 bears a note that the leaves of this species are eaten by the Karamojong, and *Raymond* 109 notes its use as spinach.

*Moquin cited Thunberg specimens from Herbs. Krauss and Hooker. The Krauss specimen has not been traced and may be B †. No Thunberg specimen from Herb. Hooker exists at Kew.

7 **A. graecizans** *L.*, Sp. Pl.: 990 (1753); F.P.S. 1: 117 (1950); Cavaco in Mém. Mus. Nat. Hist. Nat. Paris, sér. B, 13: 52 (1962); U.K.W.F.: 133 (1974). Type: material from Uppsala Botanic Garden, *Herb. Linnaeus* 1117.3 (LINN, lecto.!)

Annual herb, branched from the base and usually also above, erect, decumbent or prostrate, mostly up to 45(–70) cm. Stem slender to stout, angular, glabrous or thinly to moderately furnished with short to long, often crisped multicellular hairs which increase upwards, especially in the inflorescence. Leaves glabrous or sometimes sparingly furnished on the lower surface of the principal veins with very short gland-like hairs, long-petiolate (petiole from 3–5 mm., sometimes longer than the lamina); lamina broadly ovate or rhomboid-ovate to narrowly linear-lanceolate, 4–55 × 2–30 mm., acute to obtuse or obscurely retuse at the mucronate tip, cuneate to long-attenuate at the base. Flowers all in axillary cymose clusters, ♂ and ♀ intermixed, ♂ commonest in the upper cymes. Bracts and bracteoles narrowly lanceolate-oblong, pale and membranous, acuminate and with a pale or reddish arista formed by the excurrent green midrib; bracteoles subequalling or usually shorter than the perianth. Perianth-segments 3, all 1.5–2 mm.; ♂ lanceolate-oblong, cuspidate, pale and membranous, with a narrow green midrib excurrent in a short pale arista; ♀ lanceolate-oblong to linear-oblong, gradually to abruptly narrowed to the mucro, the midrib often bordered by a green vitta above and apparently thickened, the margins pale whitish to greenish. Stigmas 3, slender, usually pale, flexuous, ± 0.5 mm. Capsule subglobose to shortly ovoid, 2–2.25 mm., usually strongly wrinkled throughout with a very short smooth beak, exceeding the perianth, circumscissile or sometimes not, even on the same plant. Seeds shining, compressed, black, 1–1.25 mm., faintly reticulate especially towards the margin.

SYN. [*A. blitum* sensu P.O.A.C: 172 (1895); F.T.A. 6(1): 36 (1909) et auctt. mult., *non* L.]
[*A. polygamus* sensu P.O.A.C: 172 (1895); F.T.A. 6(1): 36 (1909) et auctt. mult., *non* L.]

KEY TO INFRASPECIFIC VARIANTS

Perianth-segments mucronate, mucro up to 0.25 mm., usually straight:
- Leaf-blades of main stem leaves linear-lanceolate to oblong, at least 2.5 times as long as broad .... a. subsp. **graecizans**
- Leaf-blades of main stem leaves elliptic-ovate to broadly ovate, less than 2.5 times as long as broad .... b. subsp. **silvestris**

Perianth-segments tapering, long-aristate, with awns mostly 0.3–0.75 mm., frequently somewhat divergent; leaf-blades linear or lanceolate to rhombic-spathulate ..... c. subsp. **thellungianus**

a. subsp. **graecizans**

Leaf-blade (in particular of the larger leaves of the main stem) at least 2.5 times as long as broad, oblong to linear-lanceolate. Perianth-segments mucronate, mucro up to ± 0.25 mm., usually straight.

KENYA. Northern Frontier Province: Ayangyangi swamp, May 1970, *Mathew* 6342!; Baringo District: Perkerra Irrigation Scheme, Jan. 1959, *Bogdan* 4740!; Lamu District: 7 km. E. of Garsen towards Witu, Mar. 1977, *Hooper & Townsend* 1229!

TANZANIA. Lushoto District: Mkomazi R., 3 km. NE. of Lake Manka, May 1953, *Drummond & Hemsley* 2352!; Dar es Salaam, Feb. 1874, *Hildebrandt* 1035!

DISTR. **K**1, 3, 7; **T**3, 6; in the Old World from the warmer parts of Europe through subtropical and tropical Asia to India, also scattered throughout most of Africa

HAB. As a weed of (usually irrigated) cultivated ground, on seasonally flooded sandy flats, or on waste ground; about sea-level to 1910 m.

SYN. *A. angustifolius* Lam., Encycl. Méth. 1: 115 (1783); F.P.N.A. 1: 130 (1948); Hauman in F.C.B. 2: 35 (1951); E.P.A.: 56 (1953), *nom. superfl.*
A. angustifolius Lam. var. *graecizans* (L.) Thell., Rep. Bot. Soc. Exch. Cl. 5: 306 (1918)

b. subsp. **silvestris** (Vill.) Brenan in Watsonia 4: 273 (1961). Type: *Herb. Tournefort* 1849 (P, lecto, IDC. microfiche 90.19!)

Leaf-blade (particularly of the main stem leaves) broadly ovate to rhombic-ovate or elliptic-ovate, less than 2.5 times as long as broad. Perianth-segments mucronate, the mucro rarely exceeding 0.25 mm., usually straight. Figs. 4/5. 5/1, 2.

UGANDA. Kigezi District: Kachwekano Farm, July 1949, *Purseglove* 3016!; Teso District: Serere, Aug. 1932, *Chandler* 853!; Mengo District: Entebbe, Sept. 1924, *Maitland* 119!
KENYA. S. Nyeri District: Mwea-Tebere Irrigation Exp. Station, Aug. 1958, *Bogdan* 4614!; Nairobi District: Nairobi National Park, May 1961, *Verdcourt & Polhill* 3151!; Teita District: Voi, Dec. 1961, *Polhill & Paulo* 937!
TANZANIA. Ngara, Dec. 1959, *Tanner* 4663!; Ufipa District: Milepa, Jan. 1949, *Burnett* 49/31!; Mbeya District: 17.5 km. SW. of Mbeya, May 1956, *Milne-Redhead & Taylor* 10071!; Zanzibar I., Makunduchi, Sept. 1974, *Mosha* 2122!
DISTR. **U** 1–4; **K** 1–7; **T** 1–8; **Z**; in the Old World from the warmer parts of Europe to the cooler regions of SW. Asia and NW. India, also in most parts of Africa.
HAB. Commonly as a weed of cultivated or waste places, also at forest edges, on rocky hillocks and in grazed or rough grassland; 0–1970 m.

SYN. *A. silvestris* Vill., Cat. Pl. Jard. Strasb. 111 (1807); E.P.A.: 60 (1953); Cavaco in Mém. Mus. Nat. Hist. Nat. Paris, sér. B, 13: 53 (1962)
A. graecizans L. var. *silvestris* (Vill.) Aschers. in Schweinf., Beitr. Fl. Aeth.: 176 (1867)
A. angustifolius Lam. var. *silvestris* (Vill.) Thell. in Schinz & Keller, Fl. Schweiz, ed. 4, 1: 222 (1923)
A. angustifolius Lam. subsp. *silvestris* (Vill.) Heukels, Geill. Schoolfl. voor Nederl., ed. 11: 170 (1934); Hauman in F.C.B. 2: 35 (1951)
A. parvulus Peter in F.R. Beih. 40(2), Descr.: 25 (1938). Type: Tanzania, Mbulu District, Ufiome, Bonga, *Peter* 44160 (B, holo.!)

NOTE. The two preceding subspecies do not appear to be truly sympatric, as might be inferred from literature, and are thus maintained at this rank. For example, subsp. *silvestris* is much the commonest form of *A. graecizans* in Europe, and also appears to be the representative race in the Caucasus and N. Iran; on the other hand, subsp. *graecizans* appears from material seen to be dominant in the warmer parts of SW. Asia (Jordan, Arabia, Iraq, S. Iran). In Africa, subsp. *graecizans* seems to be more frequent in W. and NE. tropical and South Africa, while subsp. *silvestris* is much commoner in E. Africa; this, however, may merely be a reflection of which subspecies was of earliest introduction in each region.

c. subsp. **thellungianus** *(Nevski) Gusev* in Bot. Zhurn. 57: 462 (1972). Type: U.S.S.R., Turkmenia, Kugitang, *Nevski* Pl. Turc. 730 (LE, holo.!)

Leaf-blade linear or lanceolate to rhombic-spathulate. Perianth-segments narrower, more tapering, long-aristate, awns mostly 0.3–0.75 mm. in length and frequently somewhat divergent (bracteoles also long-aristate). Fig. 4/12.

KENYA. Kiambu District: Kedong, Mt. Margaret Estate, June 1940, *Bally* 914!
TANZANIA. Lushoto District: Amani (cult.), *Greenway* AN. 1294!
DISTR. **K** 3; **T** 3 (cult); Botswana, Zimbabwe, South Africa, India (whence this subspecies may have been introduced to Africa)
HAB. Pasture land; 1910 m.

SYN. *A. blitum* L. var *polygonoides* Moq. in DC., Prodr. 13(2): 263 (1849). Type: India, Trichinopoly, *Herb. Wallich* 6906 (K, lecto.!)
A. thellungianus Nevski in Acta Acad. Sci. U.R.S.S. 1(4): 311 (1937)
A. thunbergii Moq. var. *grandifolius* Suesseng. in Mitt. Bot. Staats., München 1: 73 (1951). Type: Tanzania, Lushoto District, Amani, *Greenway* AN. 1294 (EA, holo.!)

NOTE. The type of *A. thunbergii* var. *grandifolius* Suesseng, is a very lush cultivated plant with the flower clusters becoming shortly and indistinctly spiciform, but forms approaching it (also cultivated) have been seen from India. It was said to be cultivated as a vegetable at Amani.

NOTE. (On species as a whole). Some African specimens of *A. graecizans* have been named by various botanists as subsp. *aschersonianus* Thell., which is said to differ from subsp. *graecizans* in the fruit lacking a definite circumscissile lid. However, as in *A. spinosus*, both definitely circumscissile and irregularly rupturing fruits have been found on many plants so determined, and in others the fruit has contained unripe seeds, so that dehiscence was not to be expected until further development had taken place. Hence, no taxonomic significance is given to this character in the present account.

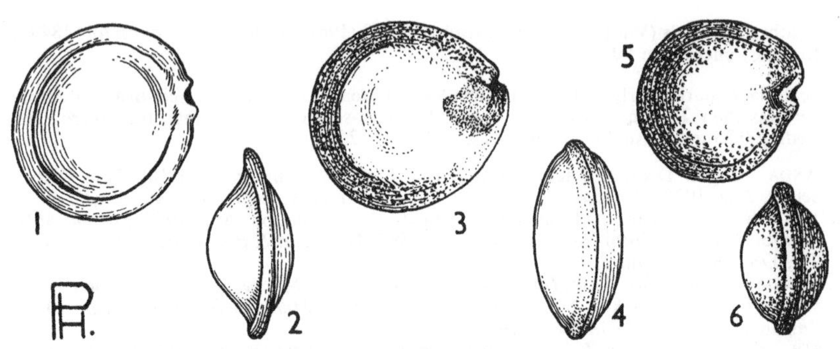

FIG. 5. Seeds of *AMARANTHUS*—**1, 2**, *A. graecizans* subsp. *sylvestris*, × 20 (*Dummer* 471); **3, 4**, *A. lividus* subsp. *polygonoides*, × 20 (*Archbold* 643); **5, 6**, *A. viridis*, × 20 (*Richards* 10900). Drawn by Pat Halliday.

8. **A. sparganiocephalus** *Thell.* in Aschers. & Graebn., Syn. Mitteleurop. Fl. 5(1): 312 (1914); Verdc. in K.B. 21: 252 (1967); E.P.A.: 59 (1953); U.K.W.F.: 133 (1974). Types: Ethiopia, Massaua, *Hildebrandt* 716 & Arabia, Chedrasch & Chedolia, *Ehrenberg* & Tanzania, 1904, *Merker* (B syn.†)

Annual herb, erect or with decumbent lower branches, 7–60 cm. Stem generally rather stout, simple or branched below the middle, yellowish, sulcate-angular, glabrous below and with rather long ± flexuous multicellular hairs especially in the flowering region, or glabrous throughout. Leaves glabrous or with scattered short hairs on the lower surface of the veins near the base, long-petiolate (petioles 0.5–5 cm. long, sometimes exceeding the lamina); lamina broadly ovate to elliptic-oblong, 1.2–4 × 0.6–3 cm., cuneate to shortly attenuate at the base, obtuse to broadly retuse at the apex. Flowers in compact spherical sessile axillary clusters to ± 1 cm. in diameter, scattered thickly along the stem and branches; ♂ and ♀ flowers intermixed. Bracts and bracteoles similar, small, oblong with a green midrib which is at most minutely percurrent; bracteoles shorter than the perianth. Perianth-segments 3, 1–1.25 mm.; ♂ elliptic-oblong and usually apiculate, ♀ oblong and obtuse; both with a central green vitta ceasing below the apex. Stigmas 2, short and rigid. Fruiting heads ± 8 mm. in diameter, rigidly stellate with the divergent capsules. Capsule 2.75–3.25 mm., rather woody, dehiscent, the persistent base conical and longitudinally sulcate, the lid subconical, also sulcate below but smooth at and below the style-base, point of junction of lid and base ± cristate-crenulate. Seeds ellipsoid, 1.25–1.5 mm., slightly compressed, black, shining, very feebly reticulate. Fig. 6.

KENYA. Northern Frontier Province: El Wak, Dec. 1959, *Joyce* in *E.A.H.* 11746!; Turkana District: Kalakol [Kaliolokoil] R., Dec. 1933, *Mortimer* T. 82!; Masai District: 21 km. on Ol Tukai-Namanga road, May 1961, *Verdcourt* 3127!
TANZANIA. Northern Province, without precise locality, 1904, *Merker*
DISTR. **K**1, 2, 6, 7; **T**2; Arabia, Aden, Socotra, Sudan, Ethiopia
HAB. Dried-up river bed, old manyatta sites, lava boulder hillside; 150–1200 m.

9. **A. deflexus** *L.*, Mant. Pl. Alt.: 295 (1771). Type: Cultivated material from Uppsala Botanic Garden, *Herb. Linnaeus* (S, lecto., IDC microfiche 384. 19!)*

Perennial herb, prostrate to somewhat ascending, (8–)12–45(–80) cm. Stem slender to rather stout, usually much branched from the base upwards, ± angular, green or reddish, glabrous below but usually increasingly furnished from well below the middle upwards with yellowish flexuous or crisped multicellular hairs. Leaves moderately to ± densely furnished on the margins and lower surface (especially of the principal veins)

*Linnean specimen No. 1117.18 (LINN), labelled as *A. scandens*, is also this species.

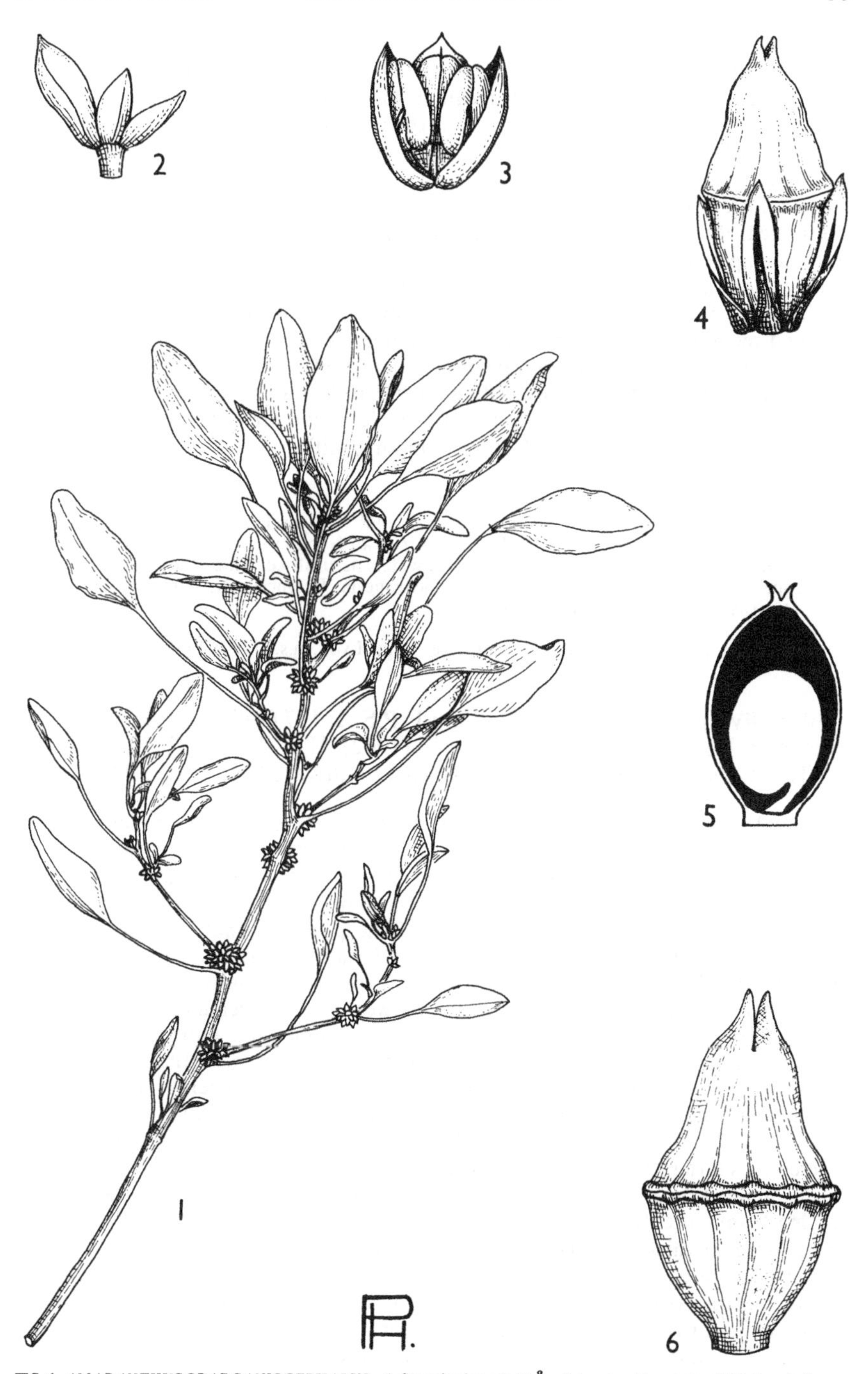

FIG. 6. *AMARANTHUS SPARGANIOCEPHALUS*—**1**, flowering branch, × $\frac{2}{3}$; **2**, bract and bracteoles, ×20; **3**, male flower, ×20; **4**, female flower, ×20; **5**, ovary, diagrammatic section showing posture of ovule, ×20; **6**, fruit, ×20. 1, from *Verdcourt* 3127; 2–6, from *Burger* 2878. Drawn by Pat Halliday.

with similar multicellular hairs, long-petiolate (petioles ± 6–25 mm. but rarely if ever exceeding the lamina); lamina broadly ovate to lanceolate (most commonly rhombic-ovate), ± 1–4.5 × 0.5–2.5 cm., subtruncate to shortly cuneate at the base, subacute to obtuse and sometimes shallowly retuse at the mucronulate apex. Flowers green, in slender and lax to stout and dense terminal and axillary spikes ± 2–10 cm. long and 5–12 mm. wide, the terminal spike not rarely with rather short stout branches and the lowest inflorescences dense subglobose clusters to ± 1 cm. in diameter; ♂ and ♀ flowers intermixed, the ♂ generally rather few. Bracts and bracteoles deltoid-ovate to ovate-lanceolate, pale and membranous with a shortly percurrent greenish midrib, bracteoles usually about half the length of the perianth. Perianth-segments 2–3, linear- to oblong-spathulate, 1.5–2 mm., obtuse to subacute, ♂ and ♀ similar or ♂ slightly more acute, pale and membranous with the thin to thick green midrib excurrent in a short paler mucro. Stigmas 2–3, pale, slender, flexuous. Capsule ellipsoid, sometimes constricted above, 1.75–3 mm., obviously exceeding the perianth, scarcely compressed, smooth, indehiscent or irregularly rupturing at maturity. Seeds compressed-ellipsoid, ± 1–1.2 × 0.7–0.8 mm., black, shining and almost smooth, with a duller slightly roughened border. Fig. 4/1.

KENYA. Nakuru, Mar. 1977, *Hooper & Townsend* 1346!; N. Nyeri District: Nanyuki, Apr. 1977, *Hooper & Townsend* 1683!; Nairobi District: Kirichwa Ndogo valley, Dec. 1951, *Bally* 8067!
DISTR. **K**3, 4; a native of temperate S. America widely naturalised in the Mediterranean region (Europe, N. Africa, Turkey), as well as in Macaronesia, Polynesia, N. America, etc.
HAB. Cultivated and waste ground; 1640–1940 m.

SYN. *Albersia deflexa* (L.) Fourr. in Ann. Soc. Linn. Lyon. n.s., 17: 142 (1869)

10. **A. lividus** *L.*, Sp. Pl.: 990 (1753); F.P.S. 1: 117 (1950); Cavaco in Mém. Mus. Nat. Hist. Nat. Paris, sér. B, 13: 55 (1962); U.K.W.F.: 132 (1974). Type-see Townsend in K.B. 29: 472 (1974): early cultivated material (BM, neo.!)

Annual herb, erect, ascending or prostrate, 6–60(–90) cm. Stem slender to stout, simple or considerably branched from the base or upwards, ± angular, green to reddish or yellow, quite glabrous or more rarely with 1–few-celled short hairs above and/or in the inflorescence. Leaves glabrous or more rarely with scattered few-celled hairs near the base on the lower surface of the primary venation, long-petiolate (petioles up to ± 10 cm., frequently longer than the lamina); lamina ovate to rhombic-ovate, 1–10 × 0.6–6 cm., shortly cuneate below, the apex usually broad and almost always distinctly emarginate, mucronulate. Flowers green, in slender to stout terminal and axillary spikes (in small forms the terminal sometimes indistinct) or rarely panicles, terminal spikes ± 0.6–11 cm. long and 0.3–2 cm. wide, or the lower axillary inflorescences of dense cymose clusters up to 2 cm. in diameter; ♂ and ♀ flowers intermixed. Bracts and bracteoles deltoid-ovate to lanceolate, whitish and membranous with a short yellow or reddish mucro formed by the excurrent midrib; bracteoles shorter than or rarely subequalling the perianth. Perianth-segments 3 (occasionally 4 or even 5 in cultivated forms), membranous margined, ♂ and ♀ both varying from lanceolate-oblong, subacute and mucronate to broadly spathulate and obtuse with the thick midrib ceasing below the summit, but the ♀ frequently blunter, 0.75–2 mm. Stigmas 2–3, short, erect or flexuous. Capsule subrotund to shortly pyriform, compressed, exceeding the perianth, 1.25–2.5 mm., usually rather smooth but sometimes wrinkled on drying, indehiscent or rupturing irregularly at maturity. Seeds 1–1.75 mm., round, compressed, dark brown to black, the centre feebly reticulate and shining, the margin duller, minutely punctate-roughened over the reticulum.

subsp. **lividus**

Plant robust to very robust, with large leaves, generally erect or ascending. Fruit 2 mm. or more in length.

UGANDA. Bunyoro, Feb. 1943, *Purseglove* 1206! (cult.); Ruwenzori, *Scott Elliott* 7941!; Mengo District: near Mpanga Forest Reserve, June 1972, *Katende* 1711!
KENYA. Trans-Nzoia District: Kitale, Nov. 1963, *Bogdan* 5654! (cult.); S. Nyeri District: Castle Forest Station, Jan. 1967, *Perdue & Kibuwa* 8394!; Kericho District: SW. Mau Forest, Mar. 1961, *Kerfoot* 2784!
TANZANIA. Lushoto District: Amani, Mar. 1941, *Greenway* 6159! (cult.); Ufipa District: N. of Sumbawanga, *Bullock* 2444!; Mbeya District: Isyesye, Mar. 1964, *Harwood* 81!
DISTR. U2, 4; K3–5; T2–4, 7; widespread in the warmer parts of Europe, east to Middle Asia, China and Japan, N. Africa from Morocco to Egypt, tropical W. and E. Africa south to Malawi, N. America
HAB. In waste ground or arable (? relic of cultivation), in forest or short turf in wet places; 1470–2060 m. (wild)

SYN. *A. blitum* L., Sp. Pl.: 990(1753), *nomen confusum*. Type: *Herb. Linnaeus* 1117.4 (LINN, lecto.!)
A. oleraceus L., Sp. Pl., ed. 2: 1403 (1763); E.P.A.: 59 (1953). Type: *Herb. Linnaeus* 1117.13 (LINN, lecto.!)
A. ascendens Lois., Not. Pl. France: 141 (1810); E.P.A.: 57 (1953). Type: Bauhin, Hist. Plant. 2: 966, fig. "*Blitum pulchrum rectum magnum rubrum*" (lecto.!)
A. oleraceus L. var. *maxima* C.B.Cl. in F.T.A. 6(1): 34 (1909). Type: Uganda, Ruwenzori, *Scott Elliot* 7941 (K, holo.!)
A. lividus L. var. *ascendens* (Lois.) Hayward & Druce, Adventive Fl. Tweedside: 177 (1919)
A. lividus L. subsp. *ascendens* (Lois.) Heukels, Geill. Schoolfl. voor Nederl., ed. 11: 169 (1934); F.P.N.A. 1: 130 (1948); Hauman in F.C.B. 2: 33 (1951)

subsp. **polygonoides** *(Moq.) Probst*, Wolladventivfl. Mitteleur.: 74 (1949). Type: Brazil, Bahia, *Salzmann* 183a (G-DC, lecto., IDC microfiche 2188.15!)

Plant smaller and neater, with smaller (rarely as much as 4 cm. long and usually less) leaves, generally prostrate to decumbent. Fruit 1.25–1.75 mm. in length. Figs. 4/2, 5/3, 4.

UGANDA. Mengo District: Kampala Botanical Garden, 11 May 1972, *Katende* 1618! & Nansana, Feb. 1973, *Katende* 1803!
KENYA. Nakuru, Mar. 1977, *Hooper & Townsend* 1347!; Naivasha, Aug. 1934, *Turner* 3516!; Fort Hall District: Thika Horticultural Research Station, Sept. 1967, *Njogu* in *E.A.H.* 13833!
TANZANIA. Bukoba, 1939, *Culwick* 9!; Moshi District: Lekuruki village, Dec. 1969, *Richards* 24891!; Lushoto District: Amani, Sept. 1929, *Greenway* 1735!
DISTR. U 4; K 3, 4; T 1, 2, 3; widespread in the warmer temperate regions and tropics of both hemispheres, in Asia from the Caucasus through India to Malaya, Java and New Guinea; Australia; in tropical Africa mostly in the east; Macaronesia, Mascarene Is., S. America from Guyana S. to Argentina; Europe and N. America, often casual
HAB. Usually in damp places by lakes and streams or in dried-out pools, open places in forest or in waste ground, also as a stunted form on lawns and drives; 890–1820 m.

SYN. *Euxolus viridis* (L.) Moq. var. *polygonoides* Moq. in DC., Prodr. 13(2): 274 (1859)
Amaranthus lividus var. *polygonoides* (Moq.) Thell. in Rep. Bot. Soc. Exch. Club 5: 574 (1920)
A. ascendens Lois. subsp. *polygonoides* (Moq.) Priszter in Ann. Sect. Horti-et Viticult. Univ. Sc. Agric. Budapest 1951, 2(2): 221 (1953)
[*A. acutilobus* sensu Agnew, U.K.W.F.: 133 (1974), *non* Uline & Bray]

NOTE. Opinions will no doubt differ as to whether these two forms should be treated as subspecies or varieties, and true judgment is probably impossible in these weedy species whose country of origin is in doubt. The preponderance of subsp. *polygonoides* in tropical regions (especially uniform in S. America, hence the choice of lectotype) and the greater frequency of var. *lividus* in the regions of older civilisations (Europe, China), with intermediates frequent only in India and China, led me after some vacillation to choose the former rank.

11. **A. viridis** *L.*, Sp. Pl., ed. 2: 1405 (1763); F.T.A. 6(1): 33 (1909); F.P.S. 1: 117 (1950). Type: *Herb. Linnaeus* 1117.15 (LINN, lecto.!)

Annual herb, erect or more rarely ascending, 10–75(–100) cm. Stem rather slender, sparingly to considerably branched, angular, glabrous or more frequently increasingly hairy upwards (especially in the inflorescence) with short or longer and rather floccose multicellular hairs. Leaves glabrous or shortly to fairly long-pilose on the lower surface of the primary or most of the venation, long-petiolate (petioles up to ± 10 cm. long and the longest commonly longer than the lamina); lamina deltoid-ovate to rhombic-oblong, 2–7 × 1.5–5.5 cm., the margins occasionally obviously sinuate, shortly cuneate to subtruncate below, obtuse and narrowly to clearly emarginate at the tip, minutely mucronate. Flowers green, in slender axillary or terminal, frequently paniculate spikes ± 2.5–12 cm. long and 2–5 mm. wide, or in the lower part of the stem in dense axillary clusters to ± 7 mm. in diameter; ♂ and ♀ flowers intermixed but the latter more numerous. Bracts and bracteoles deltoid-ovate to lanceolate-ovate, whitish and membranous with a very short pale or reddish awn formed by the excurrent green midrib; bracteoles shorter than the perianth (± 1 mm.). Perianth-segments 3(–4); ♂ oblong-oval, acute, concave, ± 1.5 mm., shorty mucronate; ♀ narrowly oblong to narrowly spathulate, finally 1.25–1.75 mm., the borders white-membranous, minutely mucronate or not, midrib green and often thickened above. Stigmas 2–3, short, erect or almost so. Capsule subglobose, 1.25–1.75 mm., not or slightly exceeding the perianth, indehiscent or rupturing irregularly, very strongly rugose throughout. Seed ± 1–1.25 mm., round, only slightly compressed, dark brown to black with an often paler thick border, ± shining, reticulate and with shallow scurfy verrucae on the reticulum, the verrucae with the shape of the areolae. Fig. 4/8, 5/5, 6.

UGANDA. Toro District, Kasenyi, *Lock* 68/280!; Mengo District: Kampala, May 1972, *Katende* 1617! & June 1972, *Katende* 1708!
KENYA. Kwale District: Diani beach, Sept. 1977, *Sturrock* 2008 B
TANZANIA. Ufipa District: Kawa R. Gorge, Feb. 1959, *Richards* 10900! & small rocky island, Malasa Archipelago, Oct. 1964, *Richards* 19215!; Dar es Salaam, 1928, *Marshall* 30!; Zanzibar I., 27 km. towards Chwaka, July 1964, *Faulkner* 3396!
DISTR. U2, 4; K7; T3, 4, 6; Z; practically cosmopolitan in the tropical and subtropical regions of the world, penetrating more widely into the temperate regions than many of its allies (as in N. and S. America and in Europe)
HAB. Weed of cultivation, along roadside, disturbed ground near dwellings, sea-shore; 0–1200 m.

SYN. *A. gracilis* Poir., Lam. Enycl. Suppl. 1: 312 (1810); Hauman in F.C.B. 2: 32 (1951); E.P.A.: 58 (1953); Cavaco in Mém. Mus. Nat. Hist. Nat. Paris, sér. B, 13: 57 (1962)

4. DIGERA

Forssk., Fl. Aegypt.-Arab.: 65 (1775)

Pseudodigera Chiov. in Atti Ist. Bot. Pavia, ser 4, 4, 7:149 (1936)

Annual herb with alternate branches and leaves. Leaves entire. Flowers small, in long-pedunculate axillary spike-like bracteate racemes, each bract subtending a sessile or subsessile partial inflorescence consisting of a central fertile flower and 2 highly modified sterile unibracteolate lateral flowers. Perianth-segments (4–)5, the outer pair firm, nervose, mucronate, opposite and sheathing the remaining flower parts, the inner segments much more delicate and hyaline. Stamens (4–)5, free or very narrowly monadelphous at the base, without intermediate pseudostaminodes; filaments filiform; anthers bilocular. Ovary with a single ovule lateral on a curved funicle, radicle descending; style filiform; stigmas 2, divergent. Fruit a hard indehiscent nutlet enclosed by the persistent perianth and falling together with the sterile flowers and bracteoles. Endosperm copious.

One species in tropical Asia and Africa.

1. **D. muricata** *(L.) Mart.* in Nov. Act. Acad. Caes. Leop.-Carol., Nat. Curios. 13(1): 285 (1826); U.K.W.F.: 133 (1974). Type: *Herb. Linnaeus* 287.6 (LINN, lecto.!)

Annual herb, (12–)15–50(–70) cm., simple or with ascending branches from near the base; stem and branches glabrous or very sparingly pilose, with pale ridges. Leaf-blade narrowly linear to broadly ovate or rarely subrotund, (1.2–)2–6(–9) × (0.2–)0.6–3(–5) cm., glabrous or the petiole and principal veins of the lower surface of the leaf spreading hairy, acute or acuminate at the apex, gradually or (in broader-leaved forms) rapidly narrowed to the base; petiole slender, in the lower leaves up to ± 5 cm., shorter in the upper leaves. Flowers glabrous, white tinged with pink to carmine or red, more rarely greenish white, in long and slender, or shorter and denser long-pedunculate axillary racemes, up to ± 30 cm. long, laxer below; peduncles slender, the lower up to ± 14 cm., both they and the inflorescence axis glabrous or sparingly spreading hairy; bracts persistent, ovate-lanceolate, acuminate, 1–2.75 mm., glabrous, membranous with a green or brownish percurrent midrib, each subtending a sessile or subsessile partial inflorescence of 3 flowers. Central flower fertile: 2 firm navicular outer perianth-segments ± 3–4.5 mm. long, oval or oblong, 3–12-nerved, acute, mucronate; the 2–3 inner segments slightly shorter, more delicate, blunt or erose, 1–3-nerved, hyaline with a darker central vitta; stamens subequalling or shorter than the style; style ± 1.5–4 mm., the 2 stigmas finally recurved. Lateral flowers appressed, 1-bracteolate, bracteoles similar in form to the bract, these flowers much reduced and increasingly so in the upper part of the spike (sometimes absent), modified into accrescent antler-shaped scales, scales with the lateral lobes narrow to broad and wing-like. Fruit subglobose, slightly compressed, 2–2.5 mm., bluntly keeled along each side, furnished throughout with verrucae or ridges, surmounted by a thick circular rim or a corona of short firm processes; style persistent.

KEY TO INFRASPECIFIC VARIANTS

Outer tepals 7–12-nerved .... subsp. **muricata**
Outer tepals 3(–5)-nerved (subsp. *trinervis*):
  Leaves glabrous:
    Lateral lobes of sterile flowers remaining subulate or narrowly deltoid .... var. **trinervis**
    Lateral lobes of sterile flowers expanding as wings around ripe nutlet (fig. 7/8) .... var. **macroptera**
  Leaves hairy on nerves beneath .... var. **patentipilosa**

subsp. **muricata**

Outer tepals closely and prominently striate with 7–12 nerves reaching to the middle or beyond, occasionally with additional shorter intermediate nerves. Lower surface of the principal leaf veins not rarely spreading-hairy near the base of the leaf, as is the petiole. Fig. 7/10.

KENYA. Mombasa, June 1923, *R.M. Graham* in *F.D.* 648! & June 1929, *R.M. Graham,* in *F.D.* 1993!

DISTR. **K** 7; Somalia, Ethiopia, Madagascar, Socotra, tropical Asia from Yemen and tropical Arabia to Afghanistan, Pakistan, India, Malaysia and Indonesia

HAB. Not recorded, but perhaps on waste ground about the town

SYN. *Achyranthes muricata* L., Sp. Pl., ed. 2: 295 (1762)
A. alternifolia L., Mant. Pl: 50 (1767). Type: *Herb. Linnaeus* 287.7 (LINN, lecto.!)
Digera arvensis Forssk., Fl. Aegypt.-Arab.: 65 (1775); F.T.A. 6(1): 29 (1909); F.D.O.-A. 2: 210 (1932). Type: *Herb. Forsskal* Cent. 3, No. 7 (C, holo.)
Desmochaeta alternifolia (L.) DC., Cat. Hort. Monsp.: 103 (1813)
Digera alternifolia (L.) Aschers. in Schweinf., Beitr. Fl. Aeth.: 180 (1867); P.O.A.C: 172 (1895); F.P.S. 1: 119 (1950); Hauman in F.C.B. 2: 36 (1951): E.P.A.: 61 (1953); Cavaco in Mém. Mus. Nat. Hist. Nat. Paris, sér. B, 13: 59 (1962)

NOTE. This is the subspecies occurring throughout tropical Asia, and was presumably introduced into Kenya. The collector notes that the leaves are eaten as a vegetable.

subsp. **trinervis** *C.C. Townsend* in K.B. 28: 141 (1973). Type: Tanzania, Lushoto/Handeni Districts, Korogwe–Handeni road, *Faulkner* 1442 (K, holo.!)

Outer tepals not closely and prominently striate, with only a midrib and 2 long lateral nerves,

FIG. 7. *DIGERA MURICATA* subsp. *TRINERVIS* var, *TRINERVIS*—**1**, flowering plant, × ½ ; **2**, longitudinal section of flower, × 8; **3**, outer tepal, × 8; **4**, inner tepal, × 8; **5**, fruiting perianth of fertile flower, × 8; **6**, fruit, × 8; **7**, sterile flower, × 10. *D. MURICATA* subsp. *TRINERVIS* var. *MACROPTERA* —**8**, fruiting perianth, × 8. *D. MURICATA* subsp. *TRINERVIS* var. *PATENTIPILOSA*—**9**, lower surface of leaf, × 4, *D. MURICATA* subsp. *MURICATA*, **10**, outer tepal, × 8, 1–4, from *Newbould* 2899; 5–7, from *Semesi* 3960; 8, from *Greenway* 9778; 9, from *Jeffery* 357; 10, from *Graham* 1993. Drawn by Lura Ripley.

occasionally with 1–2 fainter intermediate nerves. Leaves and petioles usually glabrous, leaves commonly narrower than in subsp. *muricata*.

var. **trinervis**

Two lateral lobes of the sterile flowers not much expanded, accrescent but remaining subulate or narrowly deltoid. Fig. 7/1–7.

UGANDA. Karamoja District: Oropoi valley, June-July 1930, *Liebenberg* 150! & Kangole, May 1940, *A.S. Thomas* 3439!
KENYA. Northern Frontier Province: S. Turkana, May 1970, *Mathew* 6444!; S. Kavirondo District: Lambwe valley, 1965, *Makin* 90!; Masai District: 32 km. Magadi–Nairobi, Mar. 1956, *Verdcourt* 1466!
TANZANIA. Mbulu District: Lake Manyara National Park HQ, Mto wa Mbu, Feb. 1964, *Greenway* 11188!; Lushoto District: about 3 km. WSW. of Mkomazi on road to Moshi, Mar. 1975, *Hooper & Townsend* 1026!; Morogoro District: Ulugurus, Jan. 1935, *E.M. Bruce* 543!
DISTR. **U** 1; **K** 1–7; **T** 1–3, 5, 6; Sudan, Ethiopia, Socotra
HAB. In many kinds of habitat, extremes of dry savanna and subdesert to water, deep clay and mud being recorded, perhaps commonest on disturbed and waste ground, current and abandoned cultivation, *Acacia-Commiphora* bushland, dry seasonal stream beds, also on eroded hillsides, forest and thicket edges, etc.; 150–1520 m.

SYN. *Pseudodigera pollaccii* Chiov. in Atti Ist. Bot. Pavia, ser 4, 7: 149 (1936); E.P.A.: 62 (1953); Cavaco in Mém. Mus. Nat. Hist. Nat. Paris, sér. B, 13: 50 (1962). Type: Somalia, lower Uebi Scebeli, *Ciferri* 96 (K, isolecto.!). Divergent towards subsp. *muricata*
Digera angustifolia Suesseng. in Mitt. Bot. Staats. München 1: 61 (1950). Type: Kenya, Turkana, *Pole Evans & Erens* 1609A (K, iso.!)
D. alternifolia (L.) Aschers. var. *stenophylla* Suesseng. in Mitt. Bot. Staats., München 1: 104 (1952). Type: Kenya, Masai District, Magadi, *Bally* 5221 (K, holo.!)

var. **macroptera** *C.C. Townsend* in K.B. 28: 142 (1973). Type: Kenya, Teita District, Tsavo National Park East, *Greenway* 9778 (K, holo.!)

Two lateral lobes of the sterile flowers becoming broadly oblong or rounded in fruit, strongly nerved, appearing as wings subtending the ripe nutlet. Fig. 7/8.

UGANDA. Karamoja District: Moroto, July 1954, *J. Wilson* 118!
KENYA. Turkana District: near Lodwar, July 1954, *Hemming* 277!; Teita District: Worssera, 1965, *Hucks* 135!; Tana River District: 16 km. S. of Garissa, Jan. 1961, *Lucas* 43!
TANZANIA Moshi District: near Himo railway station, Jan. 1965, *Beesley* 80!; Pangani District, July 1893, *Volkens* 459!
DISTR. **U** 1; **K** 1–4, 6, 7; **T** 2, 3; Ethiopia
HAB. In dry seasonal stream bed, on disturbed ground at edge of *Acacia* woodland, in cleared *Acacia-Commiphora* bushland; not noted from cultivated ground; 30–1400 (–2180) m.

NOTE. This variety is strikingly different from var. *trinervis* when in fruit, but apparently indistinguishable in flower, and specimens somewhat intermediate occur, notably in Ethiopia, There is no difficulty raised by var. *macroptera* in maintaining the monotypic *Pleuropteranthera* Franch. as a genus. The characters of the gynoecium (style short with a capitate or shortly bilobed stigma in *Pleuropteranthera* but long with 2 long filiform stigmas in *Digera*) and androecium (stamens quite free in *Digera*, shortly monadelphous in *Pleuropteranthera*), together with the much broader wings of the fruit in *Pleuropteranthera*, will suffice to separate these two genera.

var. **patentipilosa** *C.C. Townsend* in K.B. 28: 142 (1973). Type: Kenya, Kilifi, *Jeffrey* K. 357 (K, holo.!, EA, iso.)

Leaves ovate to lanceolate-ovate with the primary nervation on the lower surface furnished (generally ± throughout) with patent hairs. Flowers white or greenish white (the tepals green-keeled near the tips), in generally rather short and stout spikes. Fig. 7/9.

KENYA. Kilifi District: Kibarani, Mar. 1946, *Jeffrey* K. 481!; Tana River District: Kurawa, *Polhill & Paulo* 609!; Lamu District: Kiunga, *Gillespie* 44!
DISTR. **K** 7; Socotra
HAB. In current and abandoned cultivation, along roadsides, in grassland and coastal bushland; to 6 m.

NOTE. With its broad leaves and generally short spikes this variety has more the appearance of subsp. *muricata*, but the outer tepals are 3-nerved as in subsp. *trinervis* (in one specimen some outer tepals have up to 6 nerves).

The record of *D. muricata* from Nigeria (F.W.T.A., ed. 2, 1: 148, 1954) is an error, based on a misidentification of a specimen of *Pupalia micrantha* Hauman–*Jones* in *F.H.I.* 7077. This specimen was earlier stated by a slip of the pen [Townsend in K.B. 28: 143 (1973)] to be *Cyathula prostrata* (L.) Blume.

5. NEOCENTEMA

Schinz in Viert. Nat. Ges. Zürich 56: 248 (1911)

Annual or perennial herbs with alternate leaves and branches. Leaves entire. Flowers small, in long-pedunculate axillary spike-like bracteate racemes, each bract subtending a subsessile partial inflorescence consisting of 2–3 fertile flowers, of which the outer (1–)2 are subtended by a bracteolate sterile flower of 2 laterally emerging curved accrescent horn-like processes, and the other (where 2 are present) bibracteolate. Perianth-segments (4–)5, the outer pair firm, nervose, mucronate, opposite and enclosing the remaining flower parts, the inner segments delicate, hyaline. Stamens (4–)5, free or fused by a very short basal membrane, without intermediate pseudostaminodes; filaments filiform; anthers narrow, bilocular. Ovary compressed with a firm upper rim along the major axis, the single ovule lateral on a curved funicle; radicle descending; style filiform, stigmas 2, linear. Fruit 1–3 capsules enveloped by the developing mass of the processes, receptacle and very short pedicel, each partial inflorescence becoming indurate, spiky and gall-like, containing 1–3 seeds in the compressed capsules, falling as a unit. Endosperm copious.

Two species, one in East Africa, the other in Somalia.

It seems clear that the affinity of this genus is with *Digera* rather than with the other genera with which it has been associated, as pointed out by Suessenguth in K.B. 4: 476 (1946) and Townsend in K.B. 29: 469–70 (1974).

N. alternifolia *(Schinz) Schinz* in Viert. Nat. Ges. Zürich 56: 249 (1911). Type: Tanzania, S. Masai steppe, *Stuhlmann* 4287 (Z, holo.!, M, iso.!)

Annual or perennial herb (the reports of collectors vary), ± 0.3–1 m., with several stems from the base; stems usually simple, sometimes branched, procumbent or straggling, ridged, striate, furnished with white papillate multicellular hairs. Leaf-blade lanceolate to broadly ovate, 1–4.8 × 0.7–2.2 cm., acute to acuminate and mucronate at the apex, shortly cuneate to attenuate at the base, subglabrous or usually the lower surface of the midrib and principal veins with divergent whitish hairs; petiole rather slender, up to ± 1.5 cm. in the lower leaves, shortening above, ± white pilose. Flowers glabrous, white to pinkish, in slender elongated or denser long-pedunculate axillary racemes (1–)2–6 cm. long, laxer below; peduncles slender, the lower up to 11 cm., both they and the inflorescence-axis glabrous; bracts persistent, ovate-lanceolate, acuminate, ± 2.5–3 mm., glabrous or sparingly ciliate, membranous, stramineous with a brownish percurrent midrib, each subtending a partial inflorescence of 2–3 ⚥ flowers. Fertile flowers with the 2 outer perianth segments firm, elliptic-oblong, 3.5–4 mm., 3–5-nerved, acute, shortly mucronate with the excurrent midrib; (2–)3 inner segments slightly shorter, hyaline and delicate, blunt or erose, 1–3-nerved, the midrib ceasing near or well below the apex; stamens subequalling the ovary with the thickened upper margin of the latter falling away from the base of the style at an angle of ± 45°; style 1.75–2 mm., slender, the stigmas finally recurved. Capsules compressed, ± 4–4.5 mm. wide and slightly shorter, delicate below but firm at the apex, the style set in a semilunar sinus on each side of which the apex is obliquely truncated. Seed ovoid, ± 3 mm., brownish, rugulose, apparently commencing to develop within the fallen partial inflorescence, which may serve as a water source. Fig. 8.

TANZANIA. Musoma District: Serengeti, Lake Magadi to Seronera, Apr. 1961, *Greenway &*

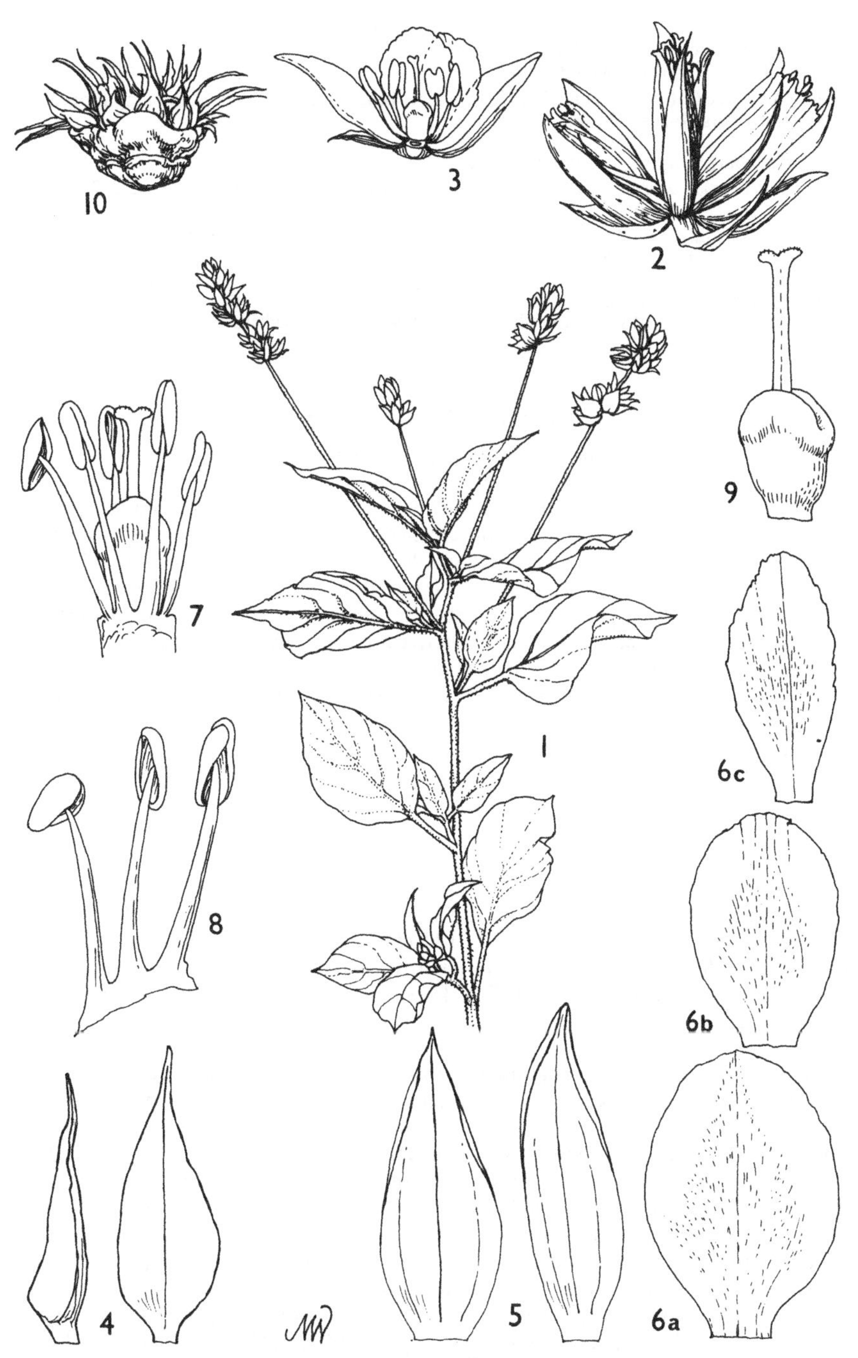

FIG. 8. *NEOCENTEMA ALTERNIFOLIA*—**1**, flowering branch, × $\frac{2}{3}$; **2**, cluster of three flowers, ×6; **3**, longitudinal section of flower, × 4; **4**, bracteoles, × 10; **5**, outer perianth segments, × 10; **6**, inner perianth segments, × 10; **7**, androecium and gynoecium, × 10; **8**, stamens, × 10; **9**, gynoecium, × 10; **10**, partial infructescence, × 2.5. All from *Richards* 22990. Drawn by Mary Millar Watt.

Myles Turner 10052!; Mbulu District: S. slopes of Mt. Hanang, Feb. 1946, *Greenway* 7750!; Kondoa District: Kinyassi, Jan. 1928, *B.D. Burtt* 979!

DISTR. **T** 1, 2, 5; not known elsewhere

HAB. As a weed of cultivation (*Sorghum*, maize, beans) on roadside and in damp ground by pools or rivers; 1430–1970 m.

SYN. *Centema alternifolia* Schinz in Bull. Herb. Boiss. 4: 419 (1896)
Sericocoma? alternifolia (Schinz) C.B. Cl. in F.T.A. 6(1): 42 (1909); F.D.O.-A. 2: 216 (1932)

NOTE. Misleadingly similar to *Digera muricata* var. *patentipilosa* in general appearance when in flower only. The 2–3-flowered partial inflorescence and the form of the ovary (strongly compressed with a linear, sloping, thickened upper edge, not a circular rim as in *Digera*) will distinguish it. In fruit it is unmistakable unless the great similarity led someone unacquainted with the present plant to assume that it was *Digera* with galled fruit!

6. SERICOSTACHYS

Gilg & Lopr. in E.J. 27: 50 (1899)

Scandent shrub with opposite branches and leaves; indumentum of jointed barbellate hairs. Leaves entire. Inflorescences a broad panicle of "spikes", with persistent bracts; partial inflorescences sessile on the spike-axis, consisting of 1 fertile bibracteolate flower and 2 modified sterile flowers, also bibracteolate. Fertile flowers ⚥ with 5 similar perianth-segments. Stamens 5, very shortly fused into a rather solid disk-like rim at the extreme base, alternating with very small pseudostaminodes; anthers bilocular. Ovary with a single pendulous ovule, glabrous, obovoid-pyriform; style filiform; stigma capitate. Sterile flowers consisting of a number of filiform hair-like appendages which become greatly elongate as fruit develops and are densely plumed with whitish hairs; bracteoles also strongly accrescent. Fruit a thin-walled indehiscent capsule enclosed by and falling with the persistent perianth and bracteoles. Endosperm copious.

A monotypic genus.

S. scandens *Gilg & Lopr.* in E.J. 27: 51 (1899) & Malpighia 14: 449 (1901); F.T.A. 6(1): 71 (1909); F.P.N.A. 1: 131 (1948); Hauman in F.C.B. 2: 70, t. 5 (1951); E.P.A.: 63 (1953); Cavaco in Mém. Mus. Nat. Hist. Nat. Paris, sér. B, 13: 62 (1962); U.K.W.F.: 134 (1974). Type: Cameroun, Yaoundé, *Zenker* 1420 (K, isolecto.!)

Much-branched scandent perennial up to at least 20 m. high, the flowering branches hanging widely from trees and, with the plumose-hairy sterile flowers, having much the appearance of a *Clematis*. Branches terete or subtetragonous, finely striate, swollen at the nodes, glabrous or more usually increasingly tomentose towards the inflorescence. Leaves broadly ovate to lanceolate-ovate, acuminate, on the main branches below the inflorescence 5.5–14×3–7 cm., subglabrous to ± densely tomentose on both surfaces, cuneate to attenuate at the base, shortly (± 1–2 cm.) petiolate. Inflorescence large, broadly paniculate with divergent branches, the branches subtended by pairs of leaves which become small and bract-like upwards, the individual "spikes" up to ± 6 × 1.5 cm., peduncles shortening above; partial inflorescences of the "spikes" sessile. Bracts deltoid-ovate, ± 2 mm., membranous with a pale midrib, glabrous or sparingly hairy; bracteoles of fertile flowers similar, 2–4 mm., bracteoles of sterile flowers at first small and closed round the rudimentary appendages, finally accrescent, lanceolate-acuminate, to ± 4–6 mm., long-pilose. Tepals 5, lanceolate to lanceolate-ovate, 3.5–8 mm., the outer 2 glabrous or floccose at the basal margin, the inner 3 ± pilose on the upper dorsal surface, all pale margined, greenish centrally with a scarcely excurrent midrib and numerous fine lateral nerves. Stamens 3.5–6 mm., very shortly connate into

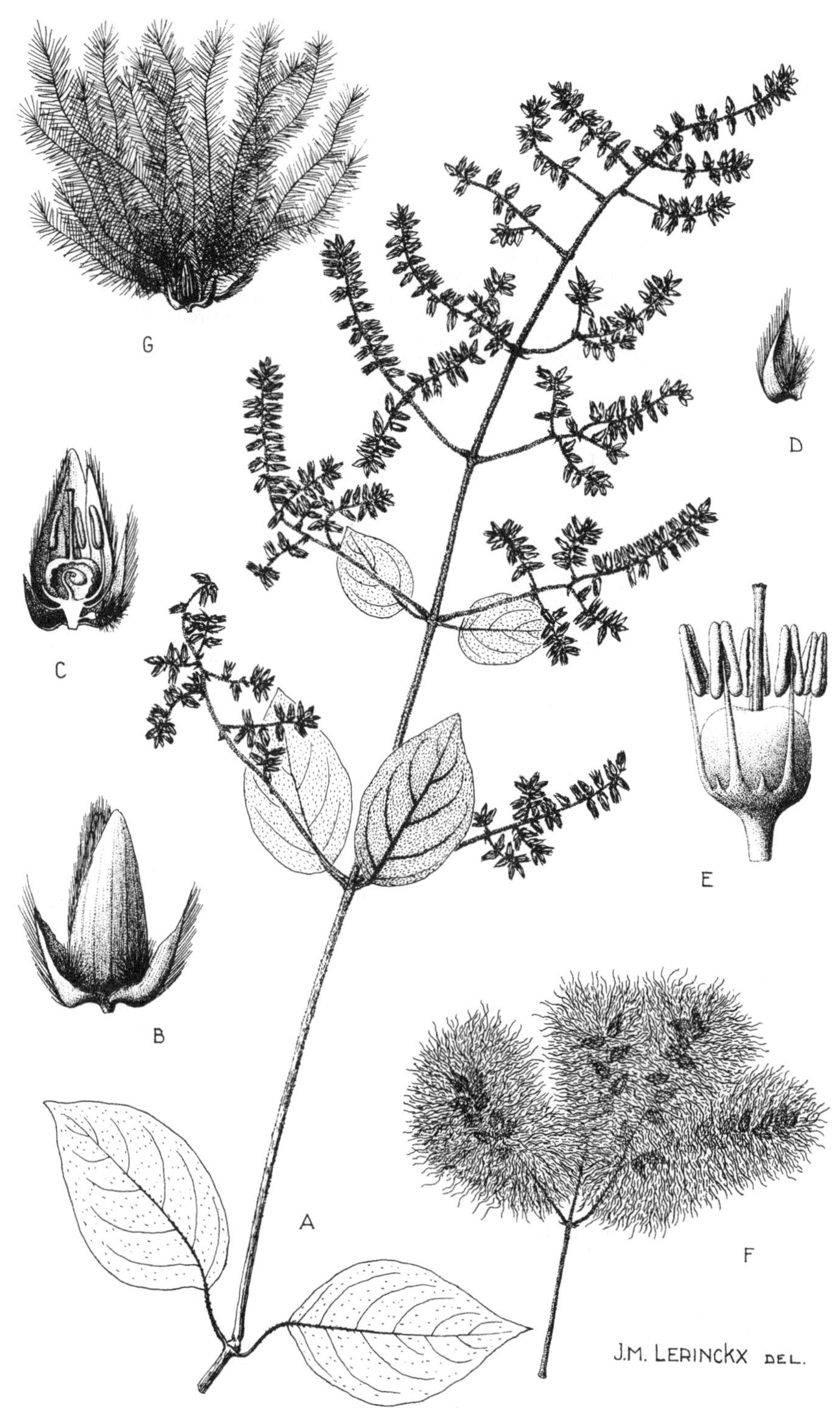

FIG. 9. *SERICOSTACHYS SCANDENS*—**A**, flowering branch, × ½ ; **B**, cymule, × 5; **C**, same in longitudinal section, × 5; **D**, bracteole with sterile flower, × 5; **E**, androecium and gynoecium, × 10; **F**, part of infructescence, × ½ ; **G**, fruiting cymule. From *Ghesquiere* 4411 and *Giorgi* 1624. Reproduced with permission from "Flore du Congo Belge".

a rim-like ring below, alternating with very small subulate or linear-oblong and toothed pseudostaminodes. Style 1.5–2.5(–3) mm. Sterile flowers of up to ± 12 linear much-accrescent appendages densely furnished with spreading jointed barbellate hairs, which much exceed the fertile perianth in fruit. Capsule ovoid-cylindrical, ± 3–3.5 mm. Seed brown, smooth and shining, ovoid, 2.5–3 mm. Fig. 9.

UGANDA. Toro District: Kivata, May 1894, *Scott Elliot* 7657!; Kigezi District: Impenetrable Forest, Mar. 1945, *Purseglove* 1634!; Mengo District: Nakabugo, Dec. 1935, *Chandler* 1486!
KENYA. Kericho District: SW. Mau Forest Reserve, 14 Aug. 1949, *Maas Geesteranus* 5789! & Kimugu Tea Estate, Dec. 1967, *Perdue & Kibuwa* 9252!; N. Kavirondo District: Kakamega Forest, Oct. 1953, *Drummond & Hemsley* 4751!
TANZANIA. Morogoro District: Uluguru Mts., N. slope of Bondwa, Oct. 1972, *Poćs & Paget-Wilkes* 6807/E!; Rungwe District: Kyimbila, Sept. 1911, *Stolz* 904!
DISTR. U2, 4; K5; T6, 7; Fernando Po, Ivory Coast, Nigeria, Cameroun, Zaire, Rwanda, Burundi, Sudan, Ethiopia, Angola, Malawi
HAB. Scrambling over tall trees or shrubs in forest (commonly riverine or lakeside); 1170–2600 m.

SYN. *S. tomentosa* Lopr. in E.J. 27: 51 (1899) & 30: 26, t. 1/P, Q (1901) & Malpighia 14: 450 (1901); F.T.A. 6(1): 71 (1909); F.D.O.-A. 2: 242 (1932); Hauman in F.C.B. 2: 72 (1951); E.P.A.: 64 (1953). Type: Uganda, Toro District, Kivata, *Scott Elliot* 7657 (K, holo.!)
S. scandens Gilg & Lopr. var. *tomentosa* (Gilg & Lopr.) Cavaco in Mém. Mus. Nat. Hist. Nat. Paris, sér. B, 13: 62 (1962)

NOTE. On two occasions [E. & P. Pf., ed. 2, 16c: 42 (1934) & Viert. Nat. Ges. Zürich 56: (1911)], Schinz wrote of his doubts as to the distinctness of *S. scandens* and *S. tomentosa*, and the present writer has no hesitation in merging them. Indumentum is generally a poor character in the Amaranthaceae – as studies in for example, *Pupalia*, *Achyranthes* and *Aerva* will soon demonstrate. West African material of the species is considerably more uniform both in indumentum and tepal size than that from East Africa.

A very odd-looking specimen from Chelima Forest, near Mafuga (U 2, Kigezi District), *Hamilton* in *Herb. Makerere Univ.* 1054 (K), has one huge galled (?) woolly head, 8 cm. in diameter, and another 4 cm. in diameter. Found named in the Kew herbarium as "*Cyathula cf. uncinulata*", it is very likely a monstrosity of *S. scandens.*

7. SERICOCOMOPSIS

Schinz in E.J. 21: 184 (May 1895) & P.O.A.C: 172 (July 1895)

Small bushy shrubs with opposite branches and leaves; indumentum of multicellular or stellate hairs. Leaves entire. Inflorescences spiciform, bracteate; flowers all ⚥ fertile, occasionally solitary but usually the partial inflorescences of 2–3 bibracteolate flowers; bracts persistent. Perianth-segments 5, similar but the inner 3 slightly shorter and narrower, all ± furnished with long barbellate hairs. Stamens 5, united below, the fertile filaments alternating with quadrate or oblong pseudostaminodes which are furnished with fimbriate dorsal scales; anthers bilocular. Ovary with a single pendulous ovule, glabrous, obpyriform; style filiform; stigma capitate. Fruit a thin-walled indehiscent capsule enclosed by the persistent perianth and bracteoles. Endosperm copious.

2 species only, both occurring in E. Africa.

Indumentum of stem and leaves simple .... 1. *S. hildebrandtii*
Indumentum of stem and leaves stellate .... 2. *S. pallida*

1. **S. hildebrandtii** *Schinz* in E.J. 21: 184 (1895) & P.O.A. C: 172 (1895) & in Viert. Nat. Ges. Zürich 56: 246 (1911); E.P.A.: 62 (1953); U.K. W.F.: 134 (1974). Type: Kenya, Teita District, Ndi, *Hildebrandt* 2584 (K, iso.!)

Much-branched bushy shrub, (0.2–)0.3–1.8 m.; cortex of old wood greyish or brownish, ridged and sometimes cracking; young branchlets densely yellowish or whitish tomentose with rather long upwardly appressed simple hairs. Leaves commonly broadly spathulate-obovate, varying to broadly ovate or oblong, more rarely narrowly obovate, or elliptical, obtuse to subacute or retuse (and then often apiculate); lamina (1–) 2–4.5 × (0.6–) 1–2.7 cm. in the larger stem leaves, shortly to longly cuneate at the base into the 0.4–0.7(–15) cm. petiole; indumentum similar to that of the branchlets, on the lower surface frequently most obvious along the midrib and primary nervation. Inflorescences 2–15(–18) × 1.4–2.5 cm., dense or lax, on a 1–3(–5) cm. peduncle, both peduncle and inflorescence-axis canescent or greyish or whitish tomentose with simple hairs; lateral cymes mostly with 2–12 flowers, flowers sometimes partly solitary in lax-flowered forms; bracts deltoid-lanceolate or deltoid-ovate, with a single excurrent midrib, 2.5–4 mm., membranous but firm, glabrous or ciliate and pilose along the midrib, persistent and deflexed; bracteoles deltoid-ovate, acuminate, ± 5–7 mm., long pilose with barbellate hairs along the apical part or commonly throughout the dorsal surface of the single excurrent midrib, frequently ciliate along the margins, arista short and straight to long, fine and ± recurved. Tepals lanceolate, 6–8 mm., the inner 3 slightly shorter, all green-vittate centrally and with hyaline margins, with a prominent midrib (excurrent in a sometimes slightly recurved arista) and 1–2 lateral nerves, densely furnished along the vitta with long barbellate hairs, or some tepals hairy in the upper part only or almost glabrous. Stamens 5–6.5 mm., the filaments flattened, alternating with oblong denticulate staminodes ± 2 mm. long, each with a dorsal fimbriate-furcate scale. Style 3.5–5.5 mm. Capsule ovoid-cylindrical, 2.5 mm., truncate with a fine rim around the apex. Seed ovoid, 2 mm., brown, smooth. Fig. 10/1–9.

UGANDA. Karamoja District: Bokora, Aug. 1957, *J. Wilson* 385!
KENYA. Northern Frontier Province: Furrole, Sept. 1952, *Gillett* 13872!; Masai District: Ol Lorgosailie, May 1943, *Bally* 2572!; Teita District: 12 km. E. of Taveta, Apr. 1975, *Friis & Hansen* 2631!
TANZANIA. Musoma District: 40 km. Seronera to Soitayai, Mar. 1961, *Greenway* 9918!; Masai District: W. of L. Natron, Mozinik, Nov. 1962, *Newbould* 6354!; Pare District: Ibaya, Mar. 1967, *Mwamba* 59!
DISTR. **U**1; **K**1–4, 6, 7; **T**1–3; S. Ethiopia
HAB. In semi-desert scrub and deciduous bushland, on rocky slopes along seasonal water-courses and near waterholes, etc.; 150–1820 m.

SYN. *Cyathula lindaviana* Lopr. in E.J. 27: 62 (1899); F.T.A. 6(1): 47 (1909), *nomen*
Cyphocarpa hildebrandtii (Schinz) C.B. Cl. in F.T.A. 6(1): 54 (1909)
Sericocomopsis lindaviana Peter in F.D.O.-A. 2: 229 (1938), *nom. illegit.* in clav. sine descr. lat.
S. meruensis Suesseng. in Mitt. Bot. Staats., München 1: 7 (1950). Type: Tanzania, Mt. Meru, *Bally* 1808 (K, holo.!)
S. grisea Suesseng. in K.B. 4: 479 (1950). Type: Tanzania, Pare District, Kiruru, *Haarer* 499 (K, holo.!, EA, iso.)

NOTE. A rather variable plant in the perianth, which may be firm and thinly pilose to more delicate and rather densely hairy, and blacken on drying or not; also in the length and direction of the awn of the bracteoles. All intermediates between the various extremes are to be found.
Glover, Gwynne & Samuel 2832 bears a note that the species is grazed by stock, and that rhinoceros are especially fond of it.

2. **S. pallida** *(S. Moore) Schinz* in E.J. 21: 185 (1895), in clav., & in Viert. Nat. Ges. Zürich 56: 246 (1911); F.D.O.-A. 2: 229 (1932); E.P.A.: 62 (1953); Cavaco in Mém. Mus. Nat. Hist. Nat. Paris, sér B, 13: 64 (1962); U.K. W.F.: 134 (1974). Type: Somalia, Meid, *Hildebrandt* 1521 (K, holo.!)

Much-branched bushy shrub, (0.2–)0.3–1.2(–1.8) m.; cortex of old wood grey to brownish, ridged and sometimes cracking, occasionally ± pruinose; young branchlets densely yellowish or whitish tomentose with short hairs stellate-branched in several planes. Leaves orbicular to ovate or broadly oval-elliptic, lamina (1.2–)2–4.5(–6.5) ×

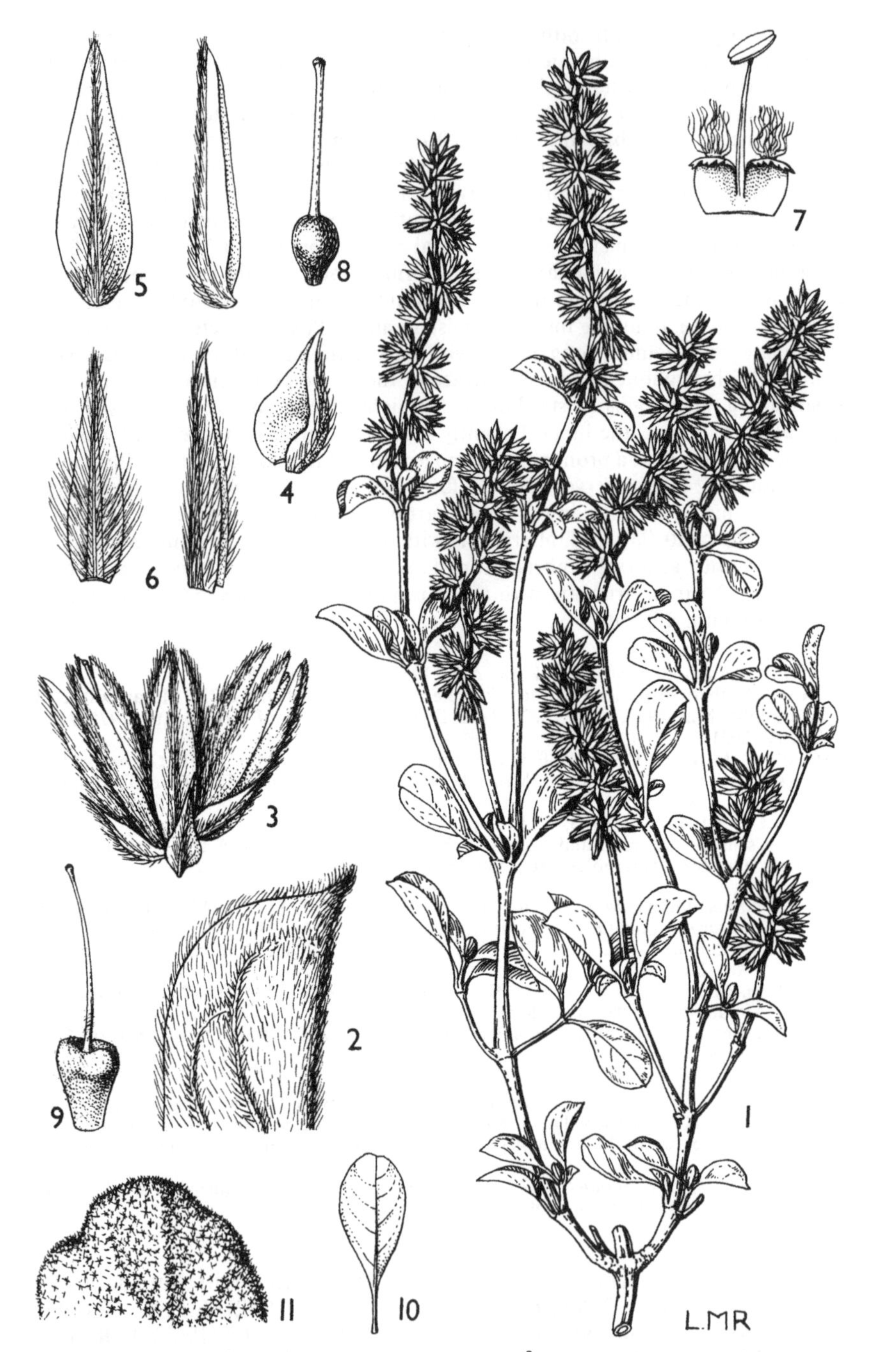

FIG. 10. *SERICOCOMOPSIS HILDEBRANDTII*—**1**, flowering branch, × $\frac{2}{3}$; **2**, leaf surface, ×4; **3**, cluster of 3 flowers, ×4; **4**, bracteole, × 6; **5**, outer tepals, × 6; **6**, inner tepals, × 6; **7**, stamen and staminodes, × 6; **8**, gynoecium, × 6; **9**, fruit, × 6. *S. PALLIDA*–**10**, leaf, × $\frac{2}{3}$; **11**, leaf-surface, × 6. 1-8, from *Richards* 25675; 9, from *Mwangangi & Gwynne* 1059; 10, 11, from *Mathenge* 53. Drawn by Lura Ripley.

(0.7–)1–3.8(–5) cm., obtuse to retuse or subacute at the apex, ± cuneate to rounded-subtruncate at the base; indumentum similar to that of the young branchlets, at first dense but later sparser and the leaves more greenish in appearance; petiole (0.2–)0.5–1.5 cm. Inflorescences (1.5–)2.5–9(–13) × 1.5–2.25 cm., dense, on a short 0.7–1.5(–3) cm. peduncle; both peduncle and inflorescence-axis stellate-tomentose; lateral cymes mostly with 4–16 flowers; bracts lanceolate-ovate, acuminate, with a single excurrent midrib, 3–4 mm., membranous, stellate-tomentose, persistent and deflexed; bracteoles deltoid-ovate or ovate, acuminate, (4–)5–7 mm., at first sparsely stellate-tomentose but soon glabrescent and shining except for jointed barbellate hairs along the single excurrent midrib and/or at the base and apex. Tepals linear, (6–)8–10 mm., the inner 3 slightly shorter, all green-vittate centrally and with hyaline margins, with a prominent excurrent midrib and 1–2 lateral nerves, densely long appressed pilose along the central vitta. Stamens ± 5 mm., the filaments flattened, alternating with oblong, subdentate pseudostaminodes ± 1.5 mm. long furnished with fimbriate-sublanate dorsal scales. Style 2–3.5 mm. Capsule ovoid, 2 mm., truncate with a fine transverse ridge. Seed ovoid, 2 mm., brown, smooth. Fig. 10/10, 11.

KENYA. Northern Frontier Province: 49 km. SW. of Ramu, Dec. 1971, *Bally & Smith* 14629!; Meru District: Meru Game Reserve, Leopard Rock Camp, June 1963, *Mathenge* 53!; Teita District: Tsavo National Park East, near Mbololo R. on the Lugard Falls road, Dec. 1966, *Greenway & Kanuri* 12635!

TANZANIA. Masai District: Longido–Moshi track, Dec. 1968, *Richards* 23545!; Pare District: NW. spur of N. Pare Mts. above Kifaru Estate, May 1968, *Bigger* 1856!; Lushoto District: Mkomazi, Apr. 1934, *Greenway* 3960!;

DISTR. **K**1, 4, 6, 7; **T**2, 3; Ethiopia, Somalia

HAB. In semi-desert scrub and deciduous bushland, rocky and disturbed places; 100–1220 m.

SYN. *Sericocoma pallida* S. Moore in J.B. 15: 70 (1877)
Cyphocarpa pallida (S. Moore) C.B. Cl. in F.T.A. 6(1): 54 (1909)
Sericocomopsis pallida (S. Moore) Schinz var. *parvifolia* Suesseng. in K.B. 4: 480 (1950), as "*parviflora*". Type: Somalia, Burao, *Glover & Gilliland* 70 (K, holo.!, EA, iso.)
S. pallida (S. Moore) Schinz var. *grandis* Suesseng. in Mitt. Bot. Staats. München 1: 343 (1953). Type: Kenya, Northern Frontier Province, Moyale, *Gillett* 13617 (K, holo.!, EA, iso.)

8. CENTEMOPSIS

Schinz in Viert. Nat. Ges. Zürich 56: 242 (1911); C.C. Townsend in Publ. Cairo Univ. Herb. 7–8: 67–72 (1977)

Annual or (? short-lived) perennial herbs with entire opposite leaves. Inflorescence terminal on the stem and branches, spiciform, capitate or fastigiate, bracteate, flowers solitary or paired in the axils of the bracts. All flowers ⚥ and bibracteolate. Bracts persistent, finally weakly deflexed or deflexed-ascending; bracteoles and perianth falling with the fruit. Perianth-segments 5, very shortly mucronate with the excurrent nerve, considerably indurate at the base in fruit. Stamens 5, the filaments delicate, shortly monadelphous at the base, alternating with distinct, quadrate or oblong, fimbriate pseudostaminodes; anthers bilocular. Ovary with a single pendulous ovule, glabrous below with a dense ring of hairs centrally and more thinly pilose above, or entirely glabrous; style slender; stigma capitate. Fruit a thin-walled capsule, irregularly ruptured by the developing seed. Seed compressed-subreniform; endosperm copious.

11 species, all tropical African.

Ovary pilose:
- Inflorescence densely cymose-fastigiate (fig. 11/5) .... 4. *C. fastigiata*
- Inflorescence a spike, solitary or paired in the axil of each bract:
  - At least some bracts subtending a pair of flowers; outer tepals 4-6 mm., sinuate-constricted about the middle in surface and lateral view especially in fruit, the hyaline margin absent or very narrow below the constriction .... 1. *C. kirkii*
  - All bracts subtending a solitary flower; outer tepals 2.5–4 mm., evenly rounded and with the hyaline margin continuous from base to apex:
    - Inflorescence dirty white, short, to 2.5 cm.; outer tepals 4 mm.; leaves thick, the margins not obviously revolute, midrib obscure .... 2. *C. sordida*
    - Inflorescence carmine to pink, more rarely whitish, usually elongate; tepals very rarely attaining 4 mm. and usually 2.5–3 mm.; leaves thinner, the margins revolute and not rarely meeting at the prominent midrib .. 3. *C. gracilenta*

Ovary glabrous:
- Inflorescence stout and very dense, ± 8–12 mm. in diameter; tepals 3–3.5 mm.; bracts narrowly oblong to narrowly spathulate, abruptly contracted to the excurrent nerve ..... 5. *C. conferta*
- Inflorescence slender, ± 3–4 mm. in diameter; tepals 1.5–2.25 mm.; bracts deltoid-ovate, ± gradually tapering to the apex:
  - Fruiting perianth falling later, so that the inflorescence becomes an elongating spike of flowers; tepals only shallowly grooved between the fine or obscure nerves, ± 2.25 mm............................. 6. *C. longipedunculata*
  - Fruiting perianths falling early, so that the inflorescence remains a cluster of flowers scarcely longer than broad with empty bracts below; tepals deeply sulcate below between the broad blunt ribs, ± 1.5 mm.......... 7. *C. filiformis*

1. **C. kirkii** *(Hook. f.) Schinz* in Viert. Nat. Ges. Zürich 56: 243 (1911) & in E. & P. Pf., ed. 2, 16C: 44 (1934); E.P.A.: 63 (1953); C.C. Townsend in Publ. Cairo Univ. Herb. 7–8: 67 (1977). Type: Malawi, W. shore of Lake Malawi [Nyasa], *Kirk* (K, lecto.!)

Annual herb (or probably short-lived perennial in more native habitats), erect, 15–100 cm. Stem wiry, strongly striate with pale ridges, smooth and glabrous except for tufts of multicellular whitish hairs about the nodes to strongly scabrid and moderately pilose, much branched from near the base upwards (simple in poorly grown forms), the branches ascending with sterile axillary short shoots. Leaves linear-filiform to lanceolate-elliptic, 1.2–8.5 cm. × 0.75–16 mm., surfaces glabrous to ± finely hairy, ± scabrid along the incrassate margins and lower surface of the midrib, gradually narrowed above to a mucronate apex, attenuate and indistinctly petiolate below. Inflorescence mauve-pink to crimson or purplish red (becoming stramineous in dried material), spiciform or reduced and capitate, 1.5–3.5 × 1–1.3 cm. in flower, rounded or conical at the tip, where more aristate bracteoles frequently give it a bristly appearance; in fruit elongating to as much as 13 cm.; axis white lanate, deeply sulcate, after the fall of the fruiting perianth often finally honeycombed with pits formed by the widening of the grooves above each flower scar, and densely clad with the persistent bracts. Bracts lanceolate to ovate-lanceolate, 2.5–5 mm., membranous with a dark shortly excurrent midrib, glabrous to ciliate. Bracteoles deltoid-ovate, 2–4 mm., acute to truncate or in the uppermost flowers sometimes even incised above, membranous

FIG. 11. *CENTEMOPSIS KIRKII* —**1**, flowering branch, × $\frac{2}{3}$; **2**, flower, opened out, ×6; **3**, gynoecium, ×6. *C. FILIFORMIS*—**4**, flowering branches, × $\frac{2}{3}$. *C. FASTIGIATA*—**5**, inflorescence, × $\frac{2}{3}$. 1-3, from *Richards* 19886; 4, from *Bax* 46; 5, from *Milne-Redhead & Taylor* 9246A. Drawn by Christine Grey-Wilson.

with a distinctly excurrent midrib, glabrous to ciliate. Flowers 1–2 in the axils of the bracts. Tepals oblong to subpanduriform, 3.5–6 mm.; outer 2 at anthesis with a deltoid-ovate, not or narrowly hyaline margined, opaque, glabrous to sparingly floccose strongly 3-nerved base, usually ± constricted above, the apical half being lanceolate, increasingly widely hyaline-margined upwards with the margin incurved and with the midrib shortly excurrent; inner 3 increasingly more delicate, narrower and more obviously constricted centrally, distinctly 3-nerved below with the hyaline margin descending much nearer the base; all tepals but (especially the outer) indurate and prominently 3-ribbed at the base in fruit (deeply sulcate between the ribs), giving the inflorescence a frequently spiky appearance. Filaments very slender, 2–3.5 mm.; pseudo-staminodes oblong or spathulate, 1–1.5 mm., denticulate to fimbriate around the apex. Ovary ovoid, glabrous and hyaline below, firm above, pilose centrally and more thinly so above; style 1.5–3.5 mm., glabrous, slightly asymmetrically placed on the ripe capsule. Capsule 1.5–2 mm., formed as the ovary but somewhat compressed. Seed compressed-ovoid, brown, ± 1.25–1.75 mm., shining, minutely reticulate. Fig. 11/1–3.

KENYA. Machakos District: Lukenya, May 1961, *Lucas & Williams* in *E.A.H.* 12323!; Masai District: Masai Mara Game Reserve, Aug. 1971, *Kokwaro & Mathenge* 2685!; Teita District: Tsavo National Park, 142 km. Arusha – Moshi – Voi road, Jan. 1962, *Greenway* 10427!
TANZANIA. Lushoto District: Mkomazi, Kisima Camp, Apr. 1967, *Mwamba* 47!; Dodoma, Apr. 1970, *Batty* 4!; Kilosa District: Great Ruaha R. valley about 20 km. SW. of Mikumi, Mar. 1975, *Hooper & Townsend* 900!
DISTR. **K**? 1, 4, 6, 7; **T**2–8;? Ethiopia, Zambia, Malawi, Mozambique, Zimbabwe, Angola
HAB. In deciduous woodland and bushland, on grassy plains, cultivated and waste ground, along roadsides, in river gullies or dried-out depressions,in rock crevices and sand pockets in lava rock pavements, always in sandy, sandy-loam or sandy-clay soil; 100–1980 m.

SYN. *Centema kirkii* Hook. f. in G.P. 3: 31 (1880); P.O.A.C: 172 (1895); F.T.A. 6(1): 56 (1909); F.D.O.-A. 2: 231 (1932)
Achyranthes breviflora Bak. in K.B. 1897: 280 (1897). Type: Malawi, between Kondowe and Karonga, *Whyte* (K, holo.!)
Centema rubra Lopr. in E.J. 27: 49 (1899) & in Malpighia 14: 442 (1901); F.D.O.–A. 2: 231 (1932); U.K.W.F.: 134 (1974). Type: Kenya, Nairobi/Masai Districts, Athi [Alhi] Plains, *Pospischil* (WU, holo.!)
[*Centemopsis biflora* sensu Schinz in E.J. 21: 183 (1895) & in Viert. Nat. Ges. Zürich 56: 242 (1911), pro parte quoad pl. or.-afr.; F.D.O.-A. 2: 231 (1932), *non* lectotypus angolensis; see Townsend in K.B. 35: 378 (1980)]
C. rubra (Lopr.) Schinz in Viert. Nat. Ges. Zürich 56: 243 (1911)
C. clausii Schinz in Viert. Nat. Ges. Zürich 57: 543 (1912). Type. Tanzania, Lushoto District, Amani, *Zimmermann* in *Herb. Amani* 2389 (EA, lecto.!)
C. kirkii (Hook. f.) Schinz forma *intermedia* Suesseng. in Mitt. Bot. Staats., München 1: 4 (1950). Type: Kenya, Masai District, Oldonyo Orok, *Bally* 4179 (K, holo.!, EA, iso.)

NOTE. The colour of the inflorescence as used in F.T.A., by Suessenguth and others to distinguish *C. kirkii* and *C. rubra* (straw-coloured in the former, red in the latter) was clearly taken on the basis of dried material only. On the basis of recent, well-annotated material, some gatherings showing the palest inflorescences in the herbarium (e.g. *Greenway & Kanuri* 14521 from Iringa District), also bear colour notes showing that they were of various shades of red when gathered; no material bearing colour notes has been described as other than red in the fresh state.

Forms with very narrow leaves and capitate inflorescences are especially frequent in rocky habitats in the Masai District of Kenya, and are no doubt xerophytic ecovars.

One specimen (*Glover, Gwynne & Samuel* 1572, from Olodolo – Gunya) bears a note that this plant is grazed by all domestic stock.

2. **C. sordida** *C.C. Townsend* in K.B. 36 : 681 (1982). Type: Kenya, Northern Frontier Province, 16 km. W. of Modo Gash, *Stannard & Gilbert* 830 (K, holo.!)

Plant up to 50 cm. tall, appearing suffruticose but duration doubtful since the type (the only known specimen) is of secondary growth after grazing or lopping. Stem wiry, tetragonous above and terete below, strongly striate with pale ridges, sparingly furnished with multicellular hairs, the branches divaricate-ascending, numerous. Leaves fasciculate, thick, linear, 12.5–30 × 1–2 mm., somewhat narrowed below, abruptly narrowed at the apex, obtuse or retuse with a yellowish mucro, glabrous (with scattered multicellular hairs when very young), lower margins sometimes with distant scabridities. Inflorescence dirty white, shortly cylindrical to subglobose, 1–2.25×0.8–1 cm.; axis densely white-lanate. Bracts ±2 mm., deltoid-ovate, hyaline and membranous with a firm, stramineous, shortly excurrent midrib, glabrous to ciliate. Bracteoles broadly oblong, ± obtuse, 3 mm., hyaline save for the shortly excurrent midrib, glabrous. Flowers solitary in the axils of the bracts. Outer 2 tepals broadly elliptic-ovate, ±4×2.5 mm., glabrous, evenly rounded from base to apex and broadly hyaline-margined throughout, the firm central area with 8–10 delicate lateral nerves in addition to the shortly excurrent midrib; inner 2 tepals spathulate-oblong, ±4×1.75 mm., the hyaline margins much wider above, the central part with 2–4 lateral nerves; third petal intermediate in form; outer tepals in particular with the midrib and 2–4 lateral nerves finally somewhat indurate at the base, slightly sulcate between these. Filaments filiform, ± 3 mm.; pseudo-staminodes ± 1.5 mm., spathulate, densely fimbriate around the apex. Ovary ovoid, glabrous and delicate below, firm above, with a median lanate band; style slender, 3 mm. long. Mature fruit unknown.

KENYA. Northern Frontier Province: Isiolo to Wajir 16 km. W. of Modo Gash, Dec. 1977, *Stannard & Gilbert* 830!

DISTR. **K**1; not known elsewhere

HAB. Open *Commiphora* bushland on very sandy soil, *Acacia nubica* present; ground flora of ephemerals, very few perennial herbs all in shelter of bushes; 390 m.

3. **C. gracilenta** *(Hiern) Schinz* in Viert. Nat. Ges. Zürich 57: 547 (1913); Hauman in F.C.B. 2: 42 (1951); Townsend in Publ. Cairo Univ. Herb. 7–8: 70 (1977). Type: Angola, Humpata plateau, *Welwitsch* 6511 (BM, iso.!)

Annual herb (or probably short-lived perennial in undisturbed ground), erect, mostly 0.5–1.6 m. Stem wiry and tough, strongly striate with pale ridges, smooth and glabrous throughout or more commonly with whitish multicellular hairs about the somewhat swollen nodes, rarely slightly scabrid on the ridges below the nodes, much branched and bushy from near the base upwards, the branches divaricate-ascending and frequently with sterile axillary short shoots. Leaves linear to linear-filiform, 1.5–9.5 cm. × 0.5–3(–4) mm., glabrous to sparingly pilose (especially when young), occasionally slightly scabrid on the pale and prominent lower surface of the midrib and on the revolute margins, sessile, sharply mucronate at the apex. Inflorescence carmine to pink or whitish, spiciform, 1.2–4 × 0.6–1 cm. in flower, and rounded or conical at the tip, elongating in fruit to as much as 20 cm.; axis very densely white-lanate, strongly furrowed but not becoming honeycombed with pits after fruit-fall, densely clad with the persistent bracts. Bracts lanceolate to ovate-lanceolate, 1.25–1.5 mm., membranous with a dark brown shortly excurrent midrib, glabrous or more commonly basally pilose and sparingly ciliate. Bracteoles deltoid-ovate, 1.25–1.75 mm., membranous with a distinctly excurrent brownish or red midrib, glabrous or sparingly pilose. Flowers solitary in the axils of the bracts. Tepals oblong, 2.5–4 mm.; outer 2 at anthesis firm centrally with a slender slightly excurrent midrib and 2 pairs of slender lateral nerves (these often branching above), glabrous, increasingly hyaline-margined upwards with the margins inflexed; inner 3 similar but slightly less firm, at least the innermost usually with only one pair of lateral nerves. Tepals slightly indurate and the outer especially somewhat more prominently 3-nerved in fruit, but never strongly ribbed as at that stage in *C. kirkii*, nor with the inflorescence appearing spiky. Filaments very slender, 1.5–3 mm., the pseudostaminodes finally ± $\frac{3}{4}$ the length of the filaments, oblong, fimbriate around the apex. Ovary ellipsoid with a dense beard of hairs around the middle, glabrous below and more thinly hairy above; style slender, 1.25–2 mm. not asymmetrically placed on the ripe capsule. Capsule pyriform, 2.25–2.5 mm., compressed. Seed compressed-ovoid, brown, 2–2.25 mm., shining, finely reticulate.

KENYA. Tana River District: S. of Garsen, June 1959, *Rawlins* 776!
TANZANIA. Ufipa District: Kasanga, old German road to Tukuyu, June 1957, *Richards* 10171!; Singida, Mar. 1928, *B.D. Burtt* 1365!; Njombe District: near Rudewa, Sept. 1970, *Thulin & Mhoro* 1206!
DISTR. **K**7; **T**4, 5, 7; Zaire, Burundi, Zambia, Malawi, Mozambique, Zimbabwe, Angola
HAB. In deciduous woodland, along roadsides, on dry shaded banks, by ditches and as a weed of cultivation – always on sandy soil; 50–2270 m.

SYN. *Centema gracilenta* Hiern, Cat. Afr. Pl. Welw. 1: 890 (1900)
Psilotrichum gracilentum (Hiern) C.B. Cl. in F.T.A. 6(1): 59 (1909)
Achyropsis laricifolia sensu Peter in F.D.O.-A. 2: 240 (1932), *nomen*
Centemopsis myurus Suesseng. in B.J.B.B. 15: 61 (1938). Type: Zambia, R. Kafue, *Lynes* 63b (BR, holo.!)
Achyropsis oxyuris Suesseng. & Overk. in Bot. Archiv. 41: 74 (1940). Type: Zambia, Mwinilunga, near source of R. Isongailu, *Milne-Redhead* 3896 (K, holo.!)

4. **C. fastigiata** *(Suesseng.) C.C. Townsend* in Publ. Cairo Univ. Herb. 7–8: 70 (1977). Type: Zaire, Shaba, Pweto–Moba (Baudouinville) road between Koshika and Kabele, *Robyns* 2143 (BR, holo.!)

Perennial herb with numerous rigidly erect stems from the base, 0.2–1.3(–2) m. tall. Stems wiry and tough, strongly striate with pale ridges, smooth and glabrous throughout or scabrid below the nodes, simple or sparingly branched with long ascending branches. Leaves narrowly linear to narrowly elliptic, 2.5–8.5 × 0.6–1.3 cm., glabrous or slightly puberulent when young, entirely smooth or scabrid along the prominent lower surface of the midrib and revolute margins, attenuate both above and below, apex sharply mucronate. Inflorescence greenish white or creamy to deep carmine-red, condensed, repeatedly branched and cymose-fastigiate, forming a ± round-topped head up to 7 cm. across but often less; shorter upper branches ± white-lanate, sulcate, after fruit-fall densely clad with the patent persistent bracts. Bracts ovate-lanceolate, ± 2 mm., membranous, mucronate with the shortly excurrent yellowish or reddish midrib, glabrous or very sparingly whitish pilose. Bracteoles similar to the bracts. Flowers solitary in the axils of the bracts. Tepals oblong, 4–5.5 mm.; outer 2 at anthesis firm centrally with a rather obscure slightly excurrent midrib and 2 faint lateral nerves, ± white pilose below, increasingly broadly hyaline-margined and delicate above with the margins inflexed; inner 3 increasingly shorter, narrow and more broadly hyaline-margined, the margins of the innermost wider than the narrow firm centre; all indurate at the base in fruit, slightly or markedly more prominently nerved. Filaments very slender, ± 5–6 mm., the pseudostaminodes ± 1–2.5 mm., oblong, fimbriate around the apex. Ovary ellipsoid, pilose in most of the upper part with the hairs sometimes ascending the base of the style at least on one side; style slender, 3.5–4 mm., symmetrically placed. Capsule ovoid-ellipsoid, 2–2.5 mm. Seed ovoid-pyriform, shining brown, feebly reticulate, ±2 mm. Fig. 11/5.

TANZANIA. Ufipa District: Namanyere, Apr. 1950, *Bullock* 2843! & Chito, Feb. 1971, *Sanane* 1529!; Songea District: about 6.5 km. W. of Songea, Mar. 1956, *Milne-Redhead & Taylor* 9246!
DISTR. **T**4, 8; Zambia and Zaire
HAB. In deciduous woodland on sand, in swampy grassland, on sandy hillside with rocky outcrop, in old cultivations; 960–1750 m.

SYN. *Robynsiella fastigiata* Suesseng. in B.J.B.B. 15: 70 (1938); Hauman in F.C.B. 2: 52 (1951); Cavaco in Mém. Mus. Nat. Hist. Nat. Paris, sér. B, 13: 155 (1962)

Centemopsis trichotoma Suesseng. in Mitt. Bot. Staats., München 1: 187 (1953). Type: Zambia, Mbala [Abercorn] to Koa, *Richards* 748 (K, holo.!)

5. **C. conferta** *(Schinz) Suesseng.* in Mitt. Bot. Staats., München 1: 187 (1953); Townsend in Publ. Cairo Univ. Herb. 7–8: 69 (1977). Type: Tanzania, Mwanza, *Stuhlmann* 4502! (Z, lecto.!)

Annual herb 0.2–1 m. tall, simple below the inflorescence or with numerous divaricate-ascending branches in the lower part of the stem. Stem and branches wiry, terete below in large plants, more usually conspicuously striate with pale ridges and green furrows, glabrous or thinly pilose in the younger parts and about the nodes. Leaves narrowly linear to filiform, 2–7 × 0.1–0.4 cm., glabrous or thinly pilose, usually scabrid at least along the revolute margins, attenuate at each end, distinctly mucronate at the apex. Inflorescences solitary and terminal or 2–4 on lateral branches, stout and very dense, spicate, (1–)1.7–8 × (0.6–)0.8–1.2 cm., greenish-white to brownish-pink or red; axis densely lanate, almost concealed by the patent-ascending dense persistent bracts. Bracts concave, narrowly oblong and almost parallel-sided to spathulate, ± 3–4 mm., membranous, abruptly narrowed to the short arista formed by the excurrent nerve, glabrous or sparingly pilose. Bracteoles narrowly lanceolate-oblong, shortly mucronate, only ± ½ as long as the perianth. Flowers solitary in the axils of the bracts, slightly compressed. Tepals oblong, 3–3.5 mm., gradually diminishing in width inwards, the 4 outer strongly 5-nerved, the innermost usually 3-nerved, all glabrous and broadly hyaline-margined above the middle, scarcely mucronate. Filaments delicate, ± 2 mm., the pseudostaminodes ± 1 mm., narrowly oblong, fimbriate at the apex. Ovary oblong-ellipsoid, glabrous; style slender, ± 1.5 mm., symmetrically placed. Capsule oblong-ellipsoid, ± 1.5 mm. Seed oblong-ellipsoid, shining brown, feebly reticulate, ± 1.25 mm.

TANZANIA. Mwanza District: Mbarika, May 1952, *Tanner* 767A!; Tabora, *Lindeman* 337!; Songea District: about 9.5 km. SW. of Songea, valley near Mtanda, June 1956, *Milne-Redhead & Taylor* 10865!

DISTR. T1, 4, 5, 7, 8; Zambia, Zaire

HAB. Grassland and light scrub, cultivated ground, cleared *Acacia, Commiphora* thickets and open *Brachystegia* woodland on hardpan and sand; 855–1180 m.

SYN. "*Amaranthaceae, novum genus*" Thomson in Speke, Journ. Source Nile, App. G:646 & 658 (1864)
Achyranthes conferta Schinz in Bull. Herb. Boiss. 4: 420 (1896)
Psilotrichum confertum (Schinz) C.B. Cl. in F.T.A. 6(1): 59 (1909)
Centrostachys conferta (Schinz) Standley in Journ. Wash. Acad. Sci. 5: 76 (1915)
Achyropsis conferta (Schinz) Schinz in E. & P. Pf., ed. 2, 16C: 63 (1934); Cavaco in Mém. Mus. Nat. Hist. Nat. Paris, sér. B, 13: 153 (1962)

6. **C. longipedunculata** *(Peter) C.C. Townsend* in Publ. Cairo Univ. Herb. 7–8: 69 (1977). Type: Tanzania, Kigoma District, N. of Lugufu on the way to Kigamba, Mwagao, *Peter* 36719 (B, holo.!)

Slender annual ± 35 cm. tall, with several divaricate-ascending branches in the lower part of the stem. Stem and branches wiry, strongly striate with pale angular ridges, quite smooth and glabrous. Leaves filiform, 10–40 × 0.5 mm., glabrous, tipped with a sharp pale mucro. Inflorescence white, spicate, 1–1.2 × 0.3–0.4 cm. (probably elongating in fruit), the axis densely white-lanate. Bracts persistent and patent, lanceolate, ± 1.5 mm., membranous, distinctly acuminate-aristate with the sharp excurrent midrib, glabrous. Bracteoles shorter and broader, deltoid-ovate, ± 1 mm., shortly mucronate. Flowers solitary in the axils of the bracts. Tepals oblong-ovate, ± 2.25 mm., all ± similar, with a broad firm faintly 3-nerved centre and rather narrow hyaline margins, shortly but rigidly mucronate at the apex. Filaments very delicate, ± 1.5 mm., the pseudostaminodes ± 1 mm., oblong, fimbriate across the apex. Ovary pyriform-ellipsoid, glabrous; style slender,±0.6 mm., symmetrically placed. Capsule ovoid-ellipsoid, 1.5 mm. Seed ovoid-ellipsoid, shining brown, feebly reticulate, 1.25 mm.

TANZANIA. Kigoma District: N. of Lugufu on way to Kigamba, Mwagao, *Peter* 36719!; Buha District: Moyowusi Controlled Area, Feb. 1972, *Mutch* 56!
DISTR. T4; Rwanda
HAB. Grassland overlying iron conglomerate, lava area in shade of taller herbs and shrubs; 1070 m.

SYN. *Pandiaka longipedunculata* Peter in F.R. Beih. 40(2), Descr.: 24, t. 31 (1938)
Achyropsis longipedunculata (Peter) Suesseng. in Bot. Archiv. 41: 85 (1940)
A. minutissima Lambinon in B.J.B.B. 47: 248 (1977). Type: Rwanda, Akagera National Park, *Robyns* 3455 (BR, holo.!)

7. **C. filiformis** *(E.A. Bruce) C.C. Townsend* in Publ. Cairo Univ. Herb. 7–8: 71 (1977). Type: Tanzania, Shinyanga, *B.D. Burtt* 2440 (K, holo.!)

Slender annual herb, much branched from the base upwards, 13–30 cm. tall, glabrous or almost so; stem and branches striate-sulcate. Leaves linear-filiform, 0.5–4 cm. long, with axillary short shoots. Inflorescences of globose to cylindrical 4–15 × 3 mm. spikes, but the flowers always appearing in a rounded head since the lower flowers and fruit fall quickly leaving only the persistent bracts; bracts lanceolate, ± 1 mm., membranous, mucronate with the slightly excurrent midrib, finally patent-deflexed; bracteoles linear-lanceolate, ± 0.75 mm. Tepals creamy- or greenish-white, lanceolate-oblong, ± 1.5 mm., rather blunt, sulcate between the midrib and the 2 lateral nerves which meet it a little above halfway, the lower dorsal surface with minute barbellate and sometimes glochidiate whitish hairs, midrib scarcely reaching the tip of the tepal. Stamens very delicate, slightly shorter than the perianth, with slender short, simple or incised, alternating pseudostaminodes. Ovary glabrous; style slender, ± 0.5 mm. Capsule ovoid, ± 1.2 mm. Seed compressed-ovoid, ± 1 mm., brown, smooth. Fig. 11/4.

TANZANIA. Mwanza, *R.L. Davis* 235!; Shinyanga, May 1931, *B.D. Burtt* 2440!; Singida District: 29 km. from Issuna on Singida–Manyoni road, Apr. 1964, *Greenway & Polhill* 11554!
DISTR. T1, 5; Burundi
HAB. In wet, grey sandy soil in marshes, glades and ground formerly cultivated; 1090–1450 m.

SYN. *Psilotrichum filiforme* E.A. Bruce in K.B. 1933: 467 (1933)

9. LOPRIOREA

Schinz in Viert. Nat. Ges. Zürich 56: 251 (1911)

Perennial herb or low subshrub with opposite leaves and branches, glabrous or almost so except in the inflorescence. Leaves entire, sessile and auriculate. Inflorescence a dense rounded or ovate head; partial inflorescences of one or two fertile flowers within the axil of each bract. Flowers bibracteolate, ⚥, the 2 oblong outer tepals much wider than the 3 linear-oblong inner tepals, all indurate at the base. Stamens 5; filaments linear, thin, without intermediate pseudostaminodes; anthers bilocular. Ovary with a single pendulous ovule, texture uniform and without a firmer apical "cap"; style filiform; stigmas very short and erect to longer and slightly divergent. Fruit an indehiscent capsule.

A monotypic genus.

L. ruspoli *(Lopr.) Schinz* in Viert. Nat. Ges. Zürich 56: 251 (1911). Type: Kenya, between Bela and Daua, *Riva (Exped. Ruspoli*) 1467 (FI, holo.!)

Sparingly to moderately branched perennial herb or low subshrub,± woody below, (5–)15–30 cm.; stem and branches terete, pale-striate, glabrous or thinly pilose. Leaves narrowly oblong to lanceolate-oblong, sessile and amplexicaul with rounded auricles, (1.5–)3–4.5 × (0.3–)0.6–1.4 cm., glabrous or thinly pilose, obtuse to subacute at the apex, mucronate. Inflorescence globose to ovoid, 0.8–1.5 cm. long and 0.8–1.25 cm. wide, on a (0.9–)1.8–4 cm. peduncle which is somewhat pilose approaching the inflorescence, axis ± densely white-hairy. Bracts broadly ovate, glabrous or ciliate, 2–3 mm., brownish-membranous with a darker vitta along the scarcely excurrent midrib, hyaline-margined. Bracteoles similar but larger, (3–)4–4.5 mm., and more broadly hyaline-margined. Flowers white. Tepals 5, the outer 2 oblong, ± 5–5.5 mm., the lower half indurate, firm, densely floccose, with 2 lateral pairs of nerves, the upper half broadly hyaline-margined to almost entirely hyaline with a dark central vitta through which passes the scarcely excurrent midrib; inner 3 similar in detail but much narrower, linear-oblong and somewhat constricted centrally, commonly with only one pair of lateral nerves, both midrib and vitta ceasing below the apex; all tepals sometimes slightly falcate. Stamens ± 4 mm. Ovary ellipsoid, glabrous, ± 1.5 mm.; style 3–4 mm., glabrous. Fruit ± 2 mm., equally firm throughout, irregularly rupturing, brownish-stramineous. Seed ± 1.75 mm., brown, ellipsoid, compressed. Fig. 12.

KENYA. Northern Frontier Province: between Bela and Daua R., June 1893, *Riva* 1467 & Ramu–Banissa road 10 km. after leaving the road to El Wak, May 1978, *Gilbert & Thulin* 1419!

DISTR. **K**1; S. Somalia, S. Ethiopia

HAB. *Commiphora, Boswellia, Acacia* bushland on stony soil over limestone; 200–500 m.

SYN. *Psilotrichum ruspolii* Lopr. in E.J. 27: 59 (1899)

10. **CYATHULA**

Blume, Bijdr. Fl. Nederl. Ind.: 548 (1826); C.C. Townsend in Publ. Cairo Univ. Herb. 7–8: 74–78 (1977), *nom. conserv.*

Annual or perennial herbs with entire opposite leaves. Inflorescence terminal on the stem and branches, spiciform or capitate, bracteate, the ultimate division basically a triad of fertile flowers, the outer pair bracteolate and subtended on the outer surface by 2 modified bracteolate flowers consisting of a number of sharply uncinately hooked (rarely glochidiate or sometimes straight) spines or bracteoliform processes, but one or both of the outer pair sometimes absent, bracteoles sometimes also with a strongly hooked arista, the spines of the modified flowers at first small, rapidly accrescent, few to many, clustered, with the clusters not or very shortly stalked. Bracts persistent, finally ± deflexed; bracteoles and perianth falling with the fruit. Perianth-segments 5, very shortly mucronate or some (especially the outer 2) hooked-aristate, serving with the bracteoles and modified flowers to distribute the fruit. Stamens 5, the filaments delicate, shortly monadelphous at the base, alternating with distinct, commonly toothed or lacerate pseudostaminodes; anthers bilocular. Ovary with a single pendulous ovule; style slender; stigma capitate. Fruit a thin-walled capsule, irregularly ruptured by the developing seed. Seed ovoid, slightly compressed; endosperm copious.

About 25 species in the tropics and subtropics of both hemispheres.

Spines of sterile modified flowers, and frequently the

FIG. 12. *LOPRIOREA RUSPOLII*—**1**, upper part of flowering plant, × $\frac{2}{3}$; **2**, bracteoles, × 8; **3**, outer tepal with hairs removed, × 8; **4**, inner tepal, × 8; **5**, perianth hairs, × c. 25; **6**, stamens, × 8; **7**, gynoecium, × 8. All from *Gilbert & Thulin* 1978. Drawn by Mary Millar Watt.

awns of their bracteoles, uncinate- or glochidiate-hooked:
Spines of sterile flowers and the awns of their bracteoles glochidiate-hooked (Fig. 13/9) .... 1. *C. braunii*
Spines of sterile flowers, and frequently the awns of their bracteoles, uncinate-hooked (Fig. 13/6):
Outer 2 tepals of the fertile flowers with the arista tipped with an uncinate hook (occasionally straight in scattered flowers of *C. uncinulata*), the hook as sharp as the hook of the bracteoles where this is present:
Flowering "spike" (and fruiting one without the spines) less than 1 cm. wide.................. 3. *C. achyranthoides*
Flowering and fruiting "spikes" at least 1 cm. wide:
Tepals glabrous .... 7. *C. uncinulata*
Tepals ± pilose with minutely barbellate hairs (Fig. 13/3)
Inner tepals narrowly lanceolate with a sharp, fine mucro, densely pilose all over or chiefly at the base; partial inflorescences (lateral cymes) ± hemispherical .... 4. *C. divulsa*
Inner tepals lanceolate-oblong, not or very obscurely mucronate, pilose chiefly about the apex; partial inflorescences ± globose (Fig. 13/1–6) .... 5. *C. polycephala*
Outer 2 tepals of the fertile flowers mucronate only, if an occasional uncinate-tipped tepal present (*C. cylindrica*), the hook not nearly as pronounced as those of the bracteoles:
Flowering "spike" (and the fruiting one without the spines) less than 1 cm. broad; bracts at most 3.5 mm., membranous but not shining and silvery-hyaline:
Outer tepals 2.25–3 mm., not or slightly exceeded by the spines of the modified flowers in fruit, the outer spines diverging to 90° or reflexed ... 2. *C. prostrata*
Outer tepals 3–5 mm., considerably exceeded by the spines in fruit, the outer spines mostly forwardly directed, rarely diverging at more than 45° .... 3. *C. achyranthoides*
Flowering "spike" 1.5–2(–2.5) cm. in diameter; bracts 7–8.5 mm., stramineous or silvery .... 6. *C. cylindrica*
Spines of sterile modified flowers and the awns of their bracteoles straight, not uncinate-hooked at the apex:
Capitula not solitary or paired, arranged in distinct racemes:
Annual with persistent leaves; burr rigidly spinous; pseudostaminodes plane or incurved at the apex, with a dentate to fimbriate dorsal scale ...... 8. *C. orthacantha*
Bushy, arching or scandent subshrub, the leaves mostly fallen at the fruiting stage; burr rather soft; pseudostaminodes dentate and plane at the apex, with no dorsal scale .... 9. *C. coriacea*
Flowers in solitary or paired, terminal, pedunculate, ± globose capitula:
Outer tepals 3.5–4 mm.; style very short, 0.3–0.5 mm.;

spines of fruiting heads considerably overtopping the fertile flowers .... 10. *C. erinacea*

Outer tepals 6.5–9 mm.; style firm, whitish, long and slender, 3.5–6 mm.; spines of fruiting heads not much overtopping the fertile flowers .... 11. *C. lanceolata*

1. **C. braunii** *Schinz* in Viert. Nat. Ges. Zürich 76: 139 (1931); C.C. Townsend in Publ. Cairo Univ. Herb. 7–8: 76 (1977). Type: Tanzania, Lindi District, Rondo–Lutamba, *Braun* in *Herb. Amani* 1276 (EA, K, Z, iso.!)

Straggling herb to ± 60 cm. Stem and branches striate, increasingly furnished above with flexuose or upwardly directed brownish multicellular hairs. Leaves broadly elliptic, ± 7 × 3 cm., acute to acuminate, cuneate at the base into a ± 5 mm. petiole, with scattered, rather long, appressed brownish multicellular hairs on both surfaces but especially along the lower surface of the primary venation. Inflorescences terminal, formed of racemes of sessile, ± distant globose heads of reduced cymes ± 1.5 cm. in diameter on an axis up to ± 10 cm. long. Bracts lanceolate-ovate, ± 6.5 mm., brownish-membranous, ± densely long-pilose dorsally with brownish multicellular hairs (these matted with the lanate hairs of the cyme axes), the rather slender midrib excurrent in a long glochidiate-tipped (barbs numerous) arista; bracteoles slightly narrower and more densely pilose; ultimate divisions of cymes of a central fertile flower and several sterile flowers. Outer 2 tepals narrowly oblong-lanceolate, 8 mm., ± densely pilose over the entire dorsal surface, narrowly hyaline-margined, 3–5-nerved with the outer 2 nerves, if present, short, the midrib excurrent in a very short glochidiate-tipped arista; inner 3 tepals similar but somewhat shorter, acute only, the midrib scarcely excurrent and not glochidiate. Modified flowers of several (to ± 12) glochidiate-tipped spines contained in 2 bracteoles smaller than those of the fertile flowers, and with the arista subequalling the lamina. Filaments slender, ± 4 mm.; pseudostaminodes cuneate-oblong, ± 2 mm., dorsally lanate-fimbriate about the incurved apex but with no dorsal scale. Ovary turbinate-obconical, ± 2 mm., with a flat, firm apex; style slender, ± 3.5 mm. Mature fruit and seeds unknown. Fig. 13/9.

KENYA. Kwale District: Lango ya Mwagandi Forest, Sept. 1968, *Gillett* 18714!
TANZANIA. Lindi District: Rondo–Lutamba, June 1906, *Braun* in *Herb. Amani* 1276!
DISTR. **K**7; **T**8; known only from the above gatherings
HAB. Forest undergrowth; only recorded altitude 360 m.

NOTE. It is remarkable that both this species and *C. ceylanica* Hook. f. (which is, as Schinz stated, clearly its closest ally) were apparently known only from the type gatherings until Gillett's recent gathering. The holotype of *C. braunii* was apparently destroyed in Berlin in World War II, and Schinz' herbarium at Zürich contains only fragments, but isotypes exist in EA and K. It is by no means impossible that if and when better material is available the Sri Lankan and E. African plants will prove to be conspecific. For the moment the somewhat larger flowers and longer style of *C. braunii* make it prudent to maintain it as distinct. I sought *C. ceylanica* unsuccessfully in Sri Lanka in 1973.

2. **C. prostrata** *(L.) Blume*, Bijdr. Fl. Nederl. Ind.: 549 (1825); P.O.A. C: 173 (1895); F.T.A. 6(1) : 43 (1909); F.D.O.-A. 2: 221 (1932); F.P.N.A. 1: 132 (1948); F.P.S. 1: 119 (1950); Hauman in F.C.B. 2: 62 (1951); E.P.A. : 65 (1953); Cavaco in Mém Mus. Nat. Hist. Nat. Paris sér. B, 13: 87 (1962); F.P.U.: 102 (1962). Type: *Herb. Linnaeus* 287.13 (LINN, lecto.!)

Annual herb (? sometimes short-lived perennial); stems prostrate and rooting at the lower nodes to erect, 0.2–1.2 m., simple or usually branched up to about the middle, ± swollen at the nodes, lower branches divaricate, the upper more erect; stem and branches bluntly 4-angled to subterete, striate or sulcate, subglabrous to ± densely pilose (especially the lower internodes). Leaves mostly rhombic to rhombic-ovate, sometimes rhombic-elliptic to shortly oval or subcircular, 1.5–8 × 1–4.5 cm., occasionally with the margin outline distinctly excavate below and/or above the middle, shortly acuminate at the apex, acute to rather blunt (more rarely rounded), shortly cuneate to cuneate-attenuate at the base, subglabrous to moderately pilose with strigose hairs on both surfaces, subsessile or distinctly (up to 13 mm.) petiolate. Spikes terminal on the stem and branches, at first dense, soon considerably elongating to as much as 35 cm. with maturing lower flowers increasingly distant, 5–7 mm. wide, peduncle up to ± 10 cm., axis and peduncle thinly to ± densely pilose; bracts and bracteoles lanceolate-ovate, ± 1.5–2 mm., mucronate with the shortly excurrent midrib, ciliate; flowers in sessile or shortly pedunculate cymose clusters (peduncles to ± 2 mm.) of 2–3⚥ flowers, the 2 laterals subtended by 2 modified flowers, or the uppermost ⚥ flowers of the spike solitary, similarly subtended by modified flowers, bibracteolate. Tepals elliptic-oblong, 2.25–3 mm., 3-nerved, subglabrous to ± densely white-pilose; outer firmer with the lateral nerves more distinct and joining the shortly excurrent midrib just below the apex, usually more densely white-pilose than the inner; inner sometimes ± falcate, slightly shorter. Spines of modified flowers sharply uncinate, numerous, glabrous, reddish, ± 2 mm., fasciculate, in fruit scarcely exceeding the tepals of the fertile flower; 2–3-flowered clusters ± globose, finally deflexed, falling as a unit to form a burr ± 5 cm. in diameter. Filaments very slender, ± 1.5 mm., the pseudostaminodes rectangular-cuneate with a truncate dentate or excavate apex. Ovary with a pileiform cap; style slender, ± 0.6 mm., often slightly swollen towards the base. Capsule ovoid, membranous save for the flat firm apex, ± 1.5 mm. Seed ovoid, smooth, ± 1.5 mm., shining, brown.

var. **prostrata**

Partial inflorescences sessile or with a short, thick peduncle not exceeding 1 mm. in length, shorter than the subtending bract. Outer tepals ± densely white-pilose. Plant frequently erect.

UGANDA. Toro District: Kirimia [Kiremia], Sept. 1932, *A.S. Thomas* 719!; Busoga District: Lolui I., May 1965, *Jackson* 15565!; Mengo District: Kyiwaga Forest, Feb. 1950, *Dawkins* 516!

TANZANIA. Bukoba District: Rwasina, May 1948, *Ford* 492!; Mpanda District: Mahali Mts., Katumba [Katimba], Sept. 1958, *Jefford & Newbould* 2340!; Morogoro District: Turiani, Nov. 1955, *Milne-Redhead & Taylor* 7415!; Zanzibar I., Kidichi, May 1961, *Faulkner* 2823!

DISTR. **U**2–4; **T**1, 3, 4, 6, 7; **Z**; practically pantropical in the Old World, S. to the Queensland rain-forests, also in Central and South America, apparently frequent in Brazil.

HAB. Usually on forest floors or in clearings, in light to dense shade, often near swamps or streams; also as a weed of disturbed areas, cultivation or roadsides; 30–1360 m.

SYN. *Achyranthes prostrata* L., Sp. Pl., ed. 2: 296 (1762)
Desmochaeta prostrata (L.) DC., Cat. Hort. Monsp.: 102 (1813)
Pupalia prostrata (L.) Mart. in Nov. Act. Acad. Caes. Leop. Carol., Nat. Curios. 13(1): 321 (1826)

var. **pedicellata** *(C.B. Cl.) Cavaco* in Fl. Cameroun 17: 46 (1974). Type: Uganda, Mengo District, Entebbe, *E. Brown* 8 (K, holo.!)

Partial inflorescences with a distinct slender peduncle up to ± 2 mm. long, equalling or exceeding the subtending bract. Outer tepals thinly white pilose to subglabrous. Plant usually prostrate and decumbent, slender.

UGANDA. Kigezi District: Malamagambo Forest, Feb. 1950, *Purseglove* 3287!; Masaka District: Sese, Bugala I., Oct. 1958, *Symes* 445!; Mengo District: Namanve Forest, Aug. 1932, *Eggeling* 879!

TANZANIA. Bukoba District: Minziro Forest Reserve, June 1958, *Procter* 937!; Lushoto District: Amani, Mar. 1975, *Hooper & Townsend* 991!; Morogoro District: Nguru Mts., Manyangu Forest, Liwale valley, Mar. 1953, *Drummond & Hemsley* 1850!

DISTR. **U**2–4; **T**1, 3, 6; widespread in tropical Africa from Sierra Leone to Ethiopia and from

Zaire to Mozambique; forms divergent to this rather weak variety occur in Indonesia, and especially New Guinea.

HAB. Usually a plant of forest floors and tracks, also in disturbed places; 600–1210 m.

SYN. *C. pedicellata* C.B. Cl. in F.T.A. 6(1): 46 (1909); F.D.O.-A. 2: 222 (1932); Cavaco in Mém. Mus. Nat. Hist. Nat. Paris, sér. B. 13: 86 (1962)
C. prostrata (L.) Blume forma *pedicellata* (C.B. Cl.) Hauman in F.C.B. 2: 64 (1951)

3. **C. achyranthoides** *(Kunth) Moq.* in DC., Prodr. 13(2): 326 (1849); P.O.A. C: 173 (1895); F.D.O.-A. 2: 224 (1932); F.P.N.A. 1: 132 (1948); Hauman in F.C.B. 2: 64 (1951); Cavaco in Mém. Mus. Nat. Hist. Nat. Paris, sér. B, 13: 85 (1962). Type: Colombia, Mompox, *Humboldt & Bonpland* (P, iso.!)

Annual (?) herb up to ± 1.2 m. tall, erect, or ascending with the lowest nodes rooting, simple or branched from the base and frequently also above; stems and branches terete near the base, becoming bluntly and finally more conspicuously tetragonous upwards, thinly to moderately pilose with appressed brownish hairs, especially about the somewhat swollen nodes. Leaves elliptic-obovate, 4–16 × 1.5–8 cm., acute to distinctly acuminate above and mucronate with the shortly excurrent midrib, attenuate to the base and sessile, with scattered appressed hairs on the surfaces and usually more densely pilose along the midrib and primary venation. Spikes terminal on the stem and branches, dense (the flowers falling early at maturity so that the spikes remain dense, rarely with remote lower flowers), ± 8 mm. in diameter and 2–10(–15) cm. long; peduncle of terminal spikes to ± 6.5 cm.; bracts deltoid-lanceolate, 1.5–2 mm., membranous, mucronate with the shortly excurrent darker midrib, almost or quite glabrous; bracteoles deltoid-ovate, 2–3 mm., hyaline, ± long-pilose, the brownish midrib excurrent in a long frequently uncinate arista; flowers in very shortly pedunculate cymose clusters of 1–2 ⚥ flowers each subtended by 2–3 modified flowers. Tepals oblong-lanceolate, 3–5 mm., 3-nerved with the lateral nerves strongest in the outer tepals, long-pilose, midrib joined by the lateral nerves just below the apex and excurrent to form a short mucro, or in one or more of the outer tepals an uncinate arista. Modified flowers of a number of uncinate-aristate processes, the outer deltoid-ovate below and bracteole-like, the remainder gradually narrower and transitional to ± 2–4 hooked spines; flower clusters ovoid, finally deflexed, falling as a unit to form a burr ± 6 mm. long with the spines much exceeding the fertile flower(s). Filaments very slender, 1.75–2 mm., delicate, the pseudostaminodes subulate, less than half the length of the filaments, entire or dentate. Style slender, 1 mm. Capsule ovoid, 2 mm., delicate save for the narrow firm apex. Seed ovoid, 1.75 mm., yellowish brown, smooth and shining.

UGANDA. Masaka District: 2 km. E. of Kayugi, May 1972, *Lye* 6933! Budo Forest, 17 May 1937, *Chandler & Hancook* 7/37!; Mengo District: Tavu I., Dec. 1949, *Dawkins* 469!

TANZANIA. Bukoba District: Katoma, *Gillman* 347!

DISTR. **U4**; **T1**; widespread in tropical Africa from Sierra Leone and the Sudan to Zaire and Madagascar; also widespread throughout tropical America

HAB. In forests as ground cover, along paths, etc., often near lakes or water holes; 1130–1180 m.

SYN. *Desmochaeta achyranthoides* Kunth in H.B.K., Nov. Gen. Sp. 2: 210 (1818)
D. densiflora Kunth in H.B.K., Nov. Gen. Sp. 2: 211 (1818). Type: Colombia, Mompox, *Humboldt & Bonland* (P, iso.!)
Achyranthes geminata Thonn. in Schumach. & Thonn., Beskr. Guin. Pl.: 138 (1827). Type: Ghana, *Thonning* 216 (not located)
Cyathula geminata (Thonn.) Moq. in DC., Prodr. 13(2): 330 (1849)

NOTE. As Hauman observes in F.C.B. 2: 66 (1951), the distinctions between this species and *C. prostrata* are absolutely clear-cut. They can in fact be sorted by eye almost as fast as specimens can be moved.

4. **C. divulsa** *Suesseng.* in F.R. 51: 196 (1942). Type: Tanzania, Ulanga District, Sali, *Schlieben* 2245 (K, iso.!)

Herb (duration unknown), erect and apparently little branched, lax in habit, 0.3–1 m.; stem and branches slender, terete and striate below to bluntly or clearly trigonous above, greenish- to purplish-brown, ± densely furnished with spreading, ascending or sometimes deflexed brownish multicellular hairs, older parts ± glabrescent; nodes distinctly swollen, stem and branches contracted above the nodes or sometimes not. Leaves large, broadly elliptical, 5.5–18 × 3.5–10 cm., acute or acuminate at the apex, at the base abruptly narrowed or cuneate into the 1–2 cm. petiole, moderately pilose to tomentose on both surfaces with the hairs longer on the venation of the lower surface. Inflorescences terminal on the stem and branches, each a lax spiciform thyrse formed of very shortly stalked flat- or cuneate-based hemispherical condensed cymes 1.5–2 cm. in diameter, the entire thyrse ± 3 cm. wide and up to 20 cm. long (but often less), the lower cymes increasingly distant in fruit; peduncle very short or up to 7 cm. long, both it and the inflorescence-axis densely brownish pilose; bracts deltoid- or lanceolate-ovate, 4–6.5 mm., membranous, densely pilose and scarcely shining, shortly aristate with the excurrent midrib; bracteoles similar but longer from the long-excurrent uncinate-tipped midrib, 7.5–8 mm.; ultimate divisions of the lateral cymes formed of a central fertile flower subtended on each side by a triad of one fertile and two lateral modified flowers, or all of the latter sometimes missing. Outer 2 tepals narrowly lanceolate, thinly to densely (and matted) long-pilose, 6.5–8 mm., both with a long uncinate awn formed by the excurrent midrib, narrowly hyaline-margined, with or without 1–2 fainter lateral nerves at the base which are evanescent below or a little above the middle of the tepal; inner 3 tepals shorter, 5.5–7 mm., narrowly lanceolate, more broadly hyaline-margined, thinly long-pilose to densely lanate, 3–5(–6)-nerved with the inner 2 nerves meeting the midrib below the apex but the outer much shorter, the midrib excurrent in a short rather fine and sharp mucro. Modified flowers of 2–4 narrowly lanceolate bracteoliform processes and 2–4 uncinate spines. Filaments delicate, 2.75–3 mm., pseudostaminodes 1–1.5 mm., broadly cuneate-obovate, thinly to densely lanate-pilose around the margins, dentate at the plane or incurved apex, with a subulate or finely fimbriate dorsal scale. Ovary obovoid, thickened above, 1.75–2 mm; style slender, 2–4 mm. Capsule shortly cylindrical, 2.5 mm., with a hardened rim around the concave apex. Seed ovoid, 2.25 mm., brown, almost smooth.

TANZANIA. Old Shinyanga, June 1931, *B.D. Burtt* 3296!; Dodoma District: Manyoni Kopje, May 1932, *B.D. Burtt* 3663!; Ulanga District: Sali Mission, May 1932, *Schlieben* 2245!
DISTR. T1, 5, 6; Zambia, Zimbabwe
HAB. Always in shade, in forests or among hillside shrubs; 1000–1275 m.

SYN. *C. schimperiana* Moq. var. *burttiana* Suesseng. in Mitt. Bot. Staats., München 1: 189 (1953). Type: Tanzania, Dodoma District, Manyoni, *B.D. Burtt* 3663 (K, holo.!)

NOTE. This species varies much, not only in the indumentum of the vegetative parts (as do so many Amaranthaceae) but also of the bracts, bracteoles, tepals and pseudostaminodes. On this account *Schlieben* 2245, the type of *C. divulsa*, has a very different appearance from the other two specimens cited above. But even in the limited amount of material available for study there is clear transition, and the structure of the flower parts is identical in all cases.

It is striking that whereas many weedy species of Amaranthaceae flower throughout most of the year, all 7 gatherings seen of this species were made in the period April-June, following the main rainy season.

Although very distinct in facies from *C. polycephala*, with a much more slender form, it is difficult to find reliable distinguishing characters to fit in a key. Given authentically named material for comparison, in the herbarium the two species can be sorted quickly.

5. **C. polycephala** *Bak.* in K.B. 1897: 278 (1897); F.T.A. 6 (1): 45 (1909); F.D.O.-A. 2: 223 (1932); E.P.A.: 65 (1953); C.C. Townsend, in Publ. Cairo.

Univ. Herb. 7–8: 75 (1977) Type: Kenya, Masai District, Kapiti [Kapti] plateau, *Thomson* (K, holo.!)

Perennial herb, erect and little branched to trailing with the lower nodes rooting, or semi-scandent or scandent, 0.6–2 m. or probably more when scrambling over tall bushes; stem and branches rather weak, terete and striate in the older parts, becoming tetragonous and finally more clearly angled and sulcate above, densely pilose with patent or deflexed, whitish or yellowish, multicellular hairs, the older parts more thinly hairy and occasionally glabrescent; nodes distinctly swollen, the stem and branches shrunken above the nodes when dry. Leaves broadly ovate to lanceolate-ovate or oblong-lanceolate, 4–10(–16)×(2.25–)3.5–7.5 cm., shortly cuneate to subcordate at the base, acute to shortly or more longly acuminate at the apex, moderately to densely furnished on the upper surface with appressed multicellular barbellate hairs, more densely hairy on the lower surface with the venation sometimes densely velutinous; petiole of larger leaves 0.5–1.7 cm. Inflorescences terminal on the stem and branches, each a spike-like thyrse formed of opposite or subopposite (occasionally clusters of more than 2) sessile or shortly (to ± 7 mm.) pedunculate globose condensed cymes 2–2.5 cm. in diameter, the entire thyrse (4–)6–20 cm. long with the lowest cymes increasingly distant; peduncle (3–)5–18 cm., both it and the inflorescence-axis densely yellowish or whitish pilose; bracts ovate or oblong-ovate, 4.5–5.5 mm., brownish or silvery membranous, whitish pilose at least along the midrib, which is excurrent in a short mucro; bracteoles ovate or deltoid-ovate, 4.5–7 mm., similarly membranous, thinly to moderately long-pilose at least along the midrib, which is excurrent into a distinct uncinate arista; ultimate divisions of lateral cymes formed of a central fertile flower subtended on each side by a triad of 1 fertile and 2 lateral modified flowers. Outer 2 tepals lanceolate-subnavicular, 5–7 mm., both with an uncinate awn formed by the excurrent midrib, 1-nerved, with long whitish barbellate hairs along the central dorsal surface; inner 3 tepals shorter, 3.5–5 mm., oblong-lanceolate (broader than those of *C. uncinulata*), with 3–4 slender nerves along the greenish centre, not or obscurely mucronate, margins hyaline, the apical half densely pilose with whitish barbellate hairs. Modified flowers of 2 lanceolate uncinate-tipped bracteoliform processes and a few uncinate spines of variable length. Filaments delicate, ± 2–3 mm.; pseudostaminodes 0.75–1 mm., cuneate-oblong, fimbriate. Ovary obovoid, ± 1 mm.; style slender, 2–3 mm. Capsule ovoid, ± 2 mm., membranous with a firm flattish top. Seed ovoid, ± 1.75 mm., brown, almost smooth. Fig. 13/1–6.

KENYA. Kitui District: Galunka, May 1902, *Kassner* 863!; Kericho District: S.W. Mau Forest Reserve, Aug. 1949, *Maas Geesteranus* 5635!; Masai District: Nasampolai [Enesambulai] valley, May 1969, *Greenway & Kanuri* 13628!

TANZANIA. Moshi District: Marangu, June 1893, *Volkens* 430!; Arusha District: Ngurdoto Natonal Park, Ngurdoto crater forest, Nov. 1966, *Richards* 21628!; Kondoa District: Kinyassi scarp forest, Feb. 1928, *B.D. Burtt* 1343!

DISTR. K3, 4–6; T2, 5; Ethiopia, Zaire, Rwanda

HAB. Usually in forest glades and at forest edges, sometimes near swampy ground, also secondary grassland and disturbed places; 1660–2630 m.

SYN. *C. cordifolia* Chiov. in Ann. Bot. Roma 9: 130 (1911); E.P.A.: 64 (1953). Type: Ethiopia, Debarek, *Chiovenda* 953 (FI, holo.!)

C. echinulata Hauman in B.J.B.B. 18: 107 (1946); F.P.N.A. 1: 133 (1948); Hauman in F.C.B. 2: 68 (1951). Type: Zaire, S. of Karisimbi, *de Witte* 2283 (BR, holo.!)

C. schimperiana Moq. var. *tomentosa* Suesseng. in Mitt. Bot. Staats., München 1: 190 (1953), excl. subvar. *subfusca* (which is *C. cylindrica* Moq.). Type: Kenya, Meru, *Hancock* 66 (K, lecto.!)

[*C. schimperiana* sensu Agnew, U.K.W.F.: 135 (1974), *non* Moq.]

NOTE. Unfortunately, in comparing his material of this plant with Baker's concept of *C. polycephala*, Hauman did not see the type of the latter. Since Baker, and later Clarke, made no mention of the indumentum of the tepals, Hauman appears to have assumed them to be glabrous. This is very far from the truth, and *C. echinulata* is certainly identical with *C. polycephala*.

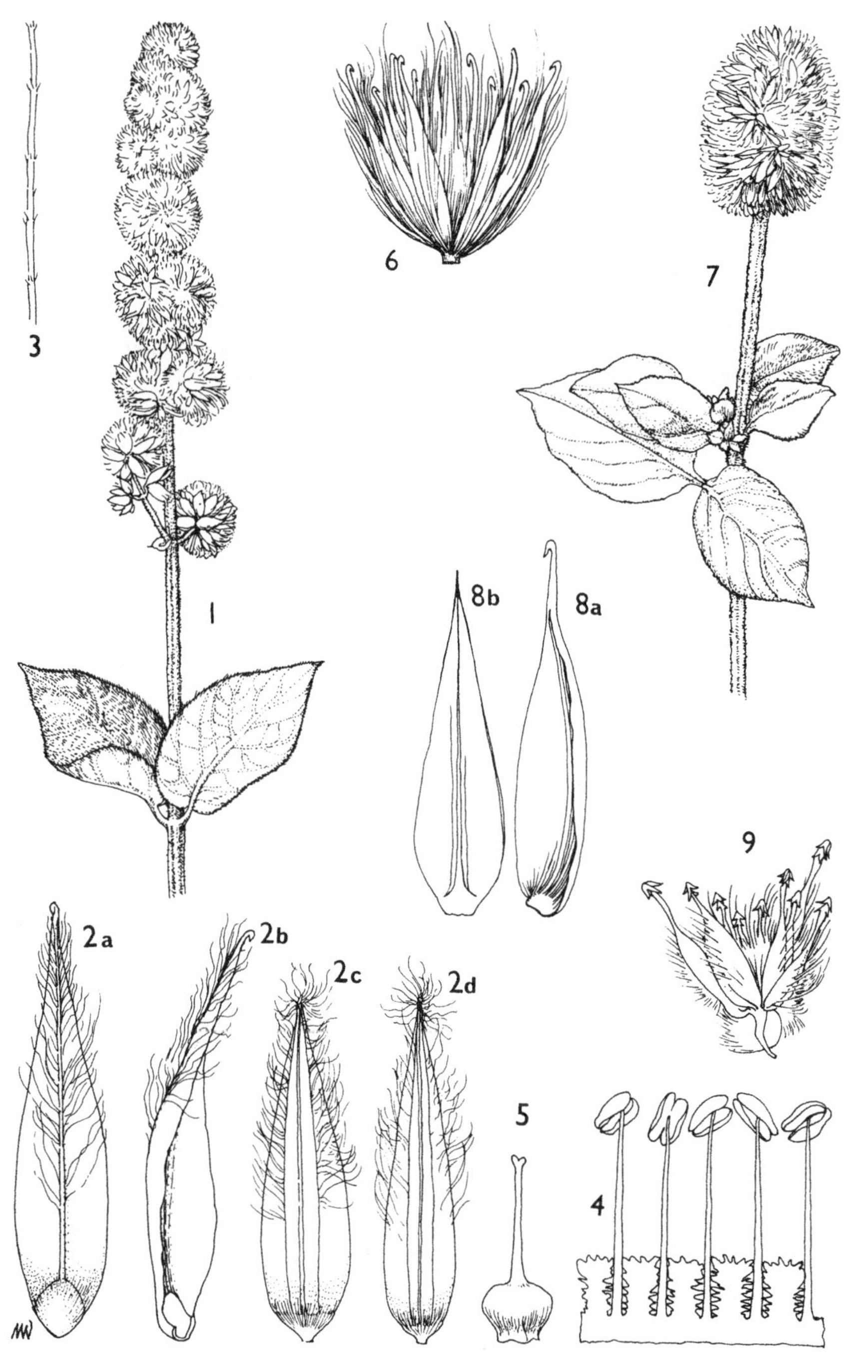

FIG. 13. *CYATHULA POLYCEPHALA*—**1**, flowering branch, × $\frac{2}{3}$; **2**, outer (a, b) inner (c, d) tepals, × 10: **3**, barbellate perianth hair, × 25; **4**, part of androecium, × 10; **5**, gynoecium, × 10; **6**, sterile flower, × 5. *C. UNCINULATA*—**7**, flowering branchlet, × $\frac{2}{3}$; **8**, outer (a) and inner (b) tepal, × 10. *C. BRAUNII*—**9**, sterile flower, × 6. 1–6, from *Gillett* 1829; 7, 8, from *Drummond* 2395; 9, from *Braun in Herb. Amani* 1275. Drawn by Mary Millar Watt.

6. **C. cylindrica** *Moq.* in DC., Prodr. 13(2); 328 (1849); P.O.A. C: 173 (1895); F.T.A. 6(1): 46 (1909); F.D.O.-A. 2: 223 (1932); F.P.N.A. 1: 132 (1948); F.P.S. 1: 119 (1950); Hauman in F.C.B. 2: 66 (1951); E.P.A.: 64 (1953); U.K. W.F.: 135 (1974); C.C. Townsend in Publ. Cairo Univ. Herb. 7-8: 74 (1977). Type: Madagascar, Tananarive, *Bojer* (G-DC, lecto., IDC microfiche 2199.7!)

Perennial herb, very variable in habit from bushy and ± 0.6–1 m. high to sprawling or decumbent and rooting at the lower nodes, or subscandent to 6 m. or more; stems and branches terete and striate in the older parts, becoming bluntly tetragonous and finally sharply tetragonous-sulcate above, glabrous or moderately to densely (especially upwards) furnished with long spreading or upwardly-appressed fuscous bristly multicellular hairs; nodes distinctly swollen in life, in dried material the stem and branches commonly shrunken just above the nodes. Leaves very variable in size and shape, small and roundish to large and broadly oblong- or elliptic-ovate, 1–14 × 0.7–6 cm., subcordate to attenuate at the base, rounded to acuminate at the apex, glabrous to ± densely furnished with long appressed multicellular hairs on both surfaces, more rarely tomentose; petiole distinct, up to ± 2.5 cm. long. Inflorescences terminal on the stem and branches, spiciform, 1.5–2(–2.5) cm. in diameter, in robust plants elongate-cylindrical and up to ± 18 cm. long, or sometimes scarcely longer than broad; "spikes" formed of densely congested (more rarely a few of the lower distant) shortly pedunculate cymose clusters composed mostly of triads of fertile flowers each subtended by 1–2 modified flowers; bracts elliptic-oblong, 7–8.5 mm., stramineous or silvery, glabrous or furnished with long multicellular hairs about the tip, aristate with the excurrent midrib, the arista usually bent but not sharply uncinate; bracteoles broadly ovate, acuminate, 4.5–9 mm., glabrous or furnished with long multicellular hairs about the tip, aristate with the excurrent midrib, the arista usually bent but not sharply uncinate. Tepals narrowly lanceolate-oblong, 4.5–7.5 mm., 3(–5)-nerved; outer 2 tepals rather feebly nerved, gibbous dorsally at the base, broadly hyaline-margined, glabrous or almost so, acute to rather blunt, the midrib excurrent in a short mucro; inner 3 tepals progressively more strongly nerved, more narrowly hyaline-margined and blunter, the innermost obtuse and often minutely lacerate-dentate at the apex with the midrib ceasing below the tip, all 3 moderately to densely furnished with long white multicellular barbellate hairs. Modified flowers with a few narrow lanceolate bracteoliform processes with uncinate tips, simple hooks, and shorter membranous scales within. Filaments slender, 3.5–5 mm., the pseudostaminodes cuneate-obovate, ± ¼ the length of the filaments, fringed above, frequently with a filiform dorsal scale. Ovary obovoid, ± 1 mm., rather firm; style slender, 2–3 mm. Capsule ovoid, 2–3 mm., membranous save for the firm apex. Seed ovoid, ± 1.75–2.75 mm., brown, shining, almost smooth.

UGANDA. Karamoja District: Mt. Moroto, Feb. 1959, *J. Wilson* 687!; Kigezi District: Kinaba Gap, Dec. 1938, *Chandler & Hancock* 2625!; Mbale District: Butandiga, Jan. 1918, *Dummer* 3680!

KENYA. Naivasha District: Njoroa Gorge [Hell's Gate], Jan. 1960, *Verdcourt* 2612!; Meru District: NE. Mt. Kenya, Sacred Lake, June 1964, *Livingstone* 64–5!; Masai District: Oloropil, May 1961, *Glover, Gwynne & Samuel* 1437!;

TANZANIA. Arusha District: Songe Hill, Feb. 1969, *Richards* 24173!; Lushoto District: Shagayu [Shagai] Forest, near Sunga, May 1953, *Drummond & Hemsley* 2623!; Mpwapwa District: Kiboriani Mt., June 1937, *Hornby & Hornby* 797!

DISTR. U1-3; K3-7; T1-7; widespread in Africa from Cameroun, Sudan and Ethiopia south to the Cape Province of South Africa

HAB. Growing in extremely diverse habitats from mist forest through *Faurea* woodland to riverside vegetation, open bushland and rocky places; 1290–2880 m.

SYN. *C. schimperiana* Moq. in DC., Prodr. 13(2): 328 (1849); P.O.A. C: 173 (1895); F.T.A. 6(1): 45 (1909); F.D.O.-A. 2: 223 (1932); E.P.A.: 65 (1953). Type: Ethiopia, Mt. Kubbi, *Schimper* I. 3 (K, iso.!)

C. mannii Bak. in K.B. 1897: 278 (1897); F.T.A. 6(1): 46 (1909); U.K.W.F.: 135 (1974). Type: Fernando Po, *Mann* 296 (K, lecto.!)

C. albida Lopr. in E.J. 27: 53 (1899). Type: Angola, Huila, *Antunes* (COI, holo.!)
Desmochaeta distorta Hiern in Cat. Afr. Pl. Welw. 1: 891 (1900). Type: Angola, Lopollo, *Welwitsch* 6487 (BM, iso.!)
Pupal huillensis Hiern in Cat. Afr. Pl. Welw. 1: 892 (1900). Type: Angola, Huila, *Welwitsch* 6493 (K, iso.!)
Cyathula distorta (Hiern) C.B. Cl. in F.T.A. 6(1): 46 (1909)
C. sp. A sensu U.K.W.F.: 135 (1974).

NOTE. Plants from extreme habitats (e.g., mist forest and dry rocky hillsides) differ much in habit and appearance. Forest plants are commonly scandent with long inflorescences and large, thinly hairy, acuminate leaves; plants from dry habitats are usually short and bushy with small, densely pilose to tomentose, ± rounded leaves. But the species varies as much in indumentum as many other species of the family (e.g. *Pupalia lappacea*), and the creation of infraspecific taxa proves equally unpractical. The numerous varieties of *C. cylindrica* and *C. schimperiana* created by Suessenguth are here treated as minor variations of no taxonomic significance.

7. **C. uncinulata** *(Schrad.) Schinz* in De Wild., Pl. Bequaert. 5: 386 (1932); F.D. O.-A. 2: 224 (1932); Hauman in F.C.B. 2: 67 (1951); E.P.A.: 65 (1953); F.F.N.R.: 45 (1962); F.P.U.: 102 (1962); U.K.W.F.: 135 (1974). Type: cultivated material from Goettingen Botanic Garden (LE, photo.!)

Erect and bushy or more commonly straggling to scandent perennial (?) herb, (0.3-)0.75-3(-6)m.; stem and branches terete and striate in the older parts, becoming bluntly tetragonous and finally sharply tetragonous-sulcate above, the older parts thinly to moderately pilose with patent or deflexed yellowish multicellular hairs, the younger parts densely pilose to thickly yellowish-tomentose or pannose; nodes distinctly swollen, when dry the stem and branches often shrunken above the nodes. Leaves broadly ovate to broadly ellipitic-oblong, 3.5-12 × 2.5-8 cm., shortly cuneate to subcordate at the base, shortly acuminate at the apex, moderately furnished on the green upper surface with appressed multicellular barbellate hairs, on the lower surface densely pilose to closely velutinous, especially along the nerves; petiole ± 1-2.25 cm. Inflorescences terminal on the stem and branches, each a dense globose head (occasionally slightly laxer, oblong and lobed) of agglomerated lateral compound cymes on a short tomentose axis, mostly 1.75-2.5 cm. in diameter when flowering but enlarging in fruit to form a vicious burr up to ± 4 cm. across, the short inflorescence branches with white lanate matted hairs; peduncle (0.6-)3-4(10) cm.; bracts deltoid-ovate, glabrous, ± 2.5 mm., stramineous, distinctly mucronate with the brown excurrent midrib; bracteoles lanceolate, ± 5-6 mm., the excurrent nerve forming a long uncinate-tipped arista; ultimate divisions of lateral cymes formed mostly of a central fertile flower subtended on each side by a triad of 1 fertile and 2 lateral modified flowers - but variable and sometimes the central fertile or one or more of lateral fertile or modified flowers absent, and occasionally the "fertile" flowers with empty anther-sacs, abortive ovary, or both. Tepals glabrous, acute, narrowly lanceolate; 2 outer tepals frequently ± transversely undulate, 5-6.5 mm., 1-nerved, the distinct arista usually uncinately hooked but sometimes not; 3 inner tepals shorter, 2.5-5 mm., 2-3-nerved with hyaline margins and a greenish centre, shortly mucronate. Modified flowers of 2 narrow uncinate-tipped bracteoliform processes and 2 shorter uncinate spines. Filaments delicate and slender, ± 2-2.5 mm.; pseudostaminodes subulate or linear-oblong to narrowly obcuneate, simple or denticulate to fimbriate, ± $\frac{1}{3}$ the length of the filaments. Ovary obovoid-turbinate, ± 1 mm., style slender, 1.75-2 mm. Capsule ovoid, ± 2-2.25 mm., membranous with a firm flattish top. Seed ovoid, ± 1.75-2 mm., brown, almost smooth. Fig. 13/7, 8.

UGANDA. Kigezi District: Kachwekano Farm, July 1949, *Purseglove* 3005!; Mbale District: Butandiga, Jan. 1918, *Dummer* 3709!; Masaka District: Kyotera, Nov. 1945, *Purseglove* 1842!
KENYA. Nyeri District: E. side of Aberdare National Park, *Hepper, Field & Townsend* 4921!; Masai District: Mara Masai Reserve, 160 km. SSW. of Narok, Sept. 1947, *Bally* 5380!; Teita District: Teita Hills, above Wusi, May 1931, *Napier* 1115!
TANZANIA. Ngara District: Bushubi, Keza, May 1960, *Tanner* 4955!; Lushoto District: Mkuzi,

May 1953, *Drummond & Hemsley* 2395!; Mbeya District: 12 km. up the Tukuyu road SW. of Mbeya, May 1956, *Milne-Redhead & Taylor* 10095!

DISTR. U1-4; K3-7; T1-8; widespread in Africa from Cameroun, Sudan and Ethiopia southward to Cape Province of South Africa, also in Madagascar

HAB. In a wide range of habitats from short grassland to open bush land and dense forest, often in the vicinity of water, sometimes in disturbed places; 900-2730 m.

SYN. *Achyranthes uncinulata* Schrad. in Ind. Sem. Hort. Goett. 1833: 1 (1833)
Cyathula globulifera Moq. in DC., Prodr. 13(2): 329 (1849); P.O.A. C: 173 (1895); F.T.A. 6(1): 44 (1909); F.D.O.-A. 2: 222 (1932); Type: Madagascar, *Bojer* (G-DC, lecto., IDC microfiche 2199.9!)
[*C. polycephala* sensu Hauman in B.J.B.B. 18: 109 (1946); F.P.N.A. 1: 133 (1948); F C.B. 2: 68 (1951), *non* Bak.]

8. **C. orthacantha** *(Aschers.) Schinz* in E. & P. Pf. 3, la: 108 (1893); P.O.A.C: 173 (1895); F.P.U.: 102 (1962); U.K.W.F.: 135 (1974). Type: Ethiopia, Goelleb and Dschadsche, *Schimper* 2153 (K, iso.!)

Annual herb, usually much-branched, erect to prostrate, commonly straggling or sprawling, 0.3–1.5 m.; stem and branches coarse, terete and striate in the lower parts, becoming bluntly tetragonous or sharply angled and sulcate above, thinly to densely furnished with white upwardly appressed or spreading multicellular hairs, the older parts usually glabrescent; nodes distinctly swollen, usually densely pilose, in larger forms the stem and branches commonly considerably shrunken above the nodes when dry. Leaves variable in form and size, from broadly ovate to broadly or narrowly elliptic, lanceolate-oblong or narrowly lanceolate, 1–15 × 0.7–5.5 cm., acute or acuminate at the apex, at the base shortly cuneate to attenuate with a petiole 0.4–2 cm. long, thinly to densely pilose with appressed white hairs which in the more densely hairy forms are longer and more divergent especially along the venation of the lower surface. Inflorescences white to pale green, crimson or carmine, terminal on the stem and branches, each a spike-like or more rarely capitate thyrse of sessile condensed cymes 4–6 cm. in diameter at anthesis, the entire thyrse 1.25–8 cm. long with the lowest cymes somewhat distant or not; peduncle 0.6–6 (–11.5) cm. long, both it and the inflorescence-axis whitish pilose; bracts lanceolate- to deltoid-ovate with a long-excurrent midrib, 3–5 mm., sparingly to moderately pilose dorsally; bracteoles broadly deltoid-ovate, 3–5 mm., midrib excurrent in a long sharp straight arista, sparingly pilose dorsally; ultimate divisions of lateral cymes formed of a cental fertile flower subtended on each side by a triad of 1 fertile and 2 lateral modified flowers. Outer 2 tepals (3.5–)4.5–7(–9) mm., with 3 very strong and prominent nerves which meet just below the apex and are excurrent to form a short mucro, usually ± densely furnished with matted multicellular barbellate white hairs, but sometimes thinly hairy or almost glabrous, and then with carmine colouration frequently developed; inner 3 tepals shorter, more faintly 3–5(–6)-nerved, with one margin usually wider below and the nerves on that side more widely separated, pilose chiefly about the apex or sometimes throughout; all tepals lanceolate-oblong, narrowly hyaline-margined. Modified flowers of a few lanceolate-based, long-aristate bracteoliform processes and several simple yellowish or reddish spines. Filaments compressed, 2.5–5 mm.; pseudostaminodes 1–2 mm., broadly cuneate-obovate, the dentate-fimbriate apex flat or incurved above, a dentate or furcate ("stags-horn") dorsal scale also present. Ovary obovoid-turbinate, ± 1–1.5 mm.; style slender, 2–5 mm. In fruit the axis and branches of the lateral cymes become indurate-incrassate and concrescent, so that each cyme falls as a complete burr 1–1.5 cm. in diameter with 4–7 mm. spines; the hard base of the burr is clad with the persistent bracteoles. Capsule pyriform, membranous save for the strongly thickened rim around the apical depression, 2.5–3 mm. Seed ovoid 2–3 mm., brown.

UGANDA. Karamoja District: Mt. Debasien, Napianyenya, Jan. 1936, *Eggeling* 2570! & 11 km. W. of Moroto, Oct. 1955, *Langdale-Brown* 1580! & Nadunget, Oct. 1958, *J. Wilson* 598!

KENYA. Northern Frontier Province: Moyale, July 1952, *Gillett* 13640!; Baringo District: 16 km. S. of L. Baringo, July 1956, *Bogdan* 4198!; Masai District: 56 km. from Nairobi on road to Magadi, July 1952, *Bogdan* 3484!
TANZANIA. Musoma District: Seronera, Apr. 1958, *Paulo* 301!; Pare District: near Same, Sept. 1969, *Batty* 588!; Kilosa District: Great Ruaha R. valley about 20 km. SW. of Mikumi, Mar. 1975, *Hooper & Townsend* 906!
DISTR. **U**1; **K**1–7; **T**1–7; Sudan, Ethiopia, Angola, Zambia, Zimbabwe, Botswana, Namibia
HAB. Occurs in many habitats, but most commonly in rough grassland with *Acacia* scrub and on disturbed and stony ground, and not rarely in the vicinity of rivers or watherholes; soil commonly sandy loam or alluvium, but also on volcanic soils and heavy black soils; 610–1460 m.

SYN. *Pupalia orthacantha* Aschers. in Schweinf., Beitr. Fl. Aeth.: 181 (1867)
Cyphocarpa orthacantha (Aschers.) C.B. Cl. in F.T.A. 6(1): 55 (1909); E.P.A.: 63 (1953)
Sericocomopsis orthacantha (Aschers.) Peter in F.D.O.-A.: 230 (1932)
Cyathula kilimandscharica Suesseng. & Beyerle in F.R. 44: 44 (1938). Type: Kenya, Masai District, Oloitokitok, *Schlieben* 5122 (B, holo. †)
Pupalia erecta Suesseng. in F.R. 44: 47 (1938). Type: Tanzania, between Dodoma and Iringa, *Troll* 5212 (B, holo. † , iso.!)
Cyathula orthacanthoides Suesseng. in Mitt. Bot. Staats., München 1: 4 (1950). Type: Tanzania, Mpwapwa, *van Rensburg* 28 (K, holo.!, EA, iso.)

NOTE. Although the type of *Cyathula kilimandscharica* is believed destroyed in Berlin in World War II, the original description agrees entirely with *C. orthacantha*, which was found in quantity at Oloitokitok in 1977 by Miss S.S. Hooper and myself.

9. **C. coriacea** *Schinz* in Viert. Nat. Ges., Zürich 76: 140 (1931); C.C. Townsend in Publ. Cairo Univ. Herb. 7–8: 75 (1977). Type: Kenya, Tana River District, Hololo, *F. Thomas* 79 (BM, Z, iso.!)

Erect and bushy to overarching or scandent subshrub, to ± 3 m. tall; stem and divergent branches closely striate, thinly to moderately furnished with appressed whitish hairs or glabrescent with age, mostly terete but sometimes a few internodes rather bluntly tetragonous; nodes scarcely swollen, the stem and branches not at all shrunken above them. Leaves narrowly to broadly oblong, elliptic or broadly ovate, 2.5–7 × 1.5–5 cm., acute to rather blunt at the apex, cuneate to truncate or subcordate at the base with a 3–12 mm. petiole, when young densely appressed-pilose on both surfaces, but soon becoming thinly hairy, falling early and few remaining on the fruiting plant. Inflorescences terminal on the stem and branches, formed of a raceme of usually opposite sessile or shortly (to ± 6 mm.) pedunculate globose finally increasingly distant heads of congested cymes, heads 1–1.25 cm. in diameter in flower, in fruit enlarging and forming burr up to 2.25 cm. across, the spines of the modified flowers and the cyme branches however not greatly accrescent or rigid and the burr much softer than that of *C. erinacea*; peduncle 2.5–7 mm., both it and the inflorescence-axis thinly to densely white pilose; bracts deltoid-ovate, 3.5–4 mm., whitish-membranous, aristate with the excurrent pale midrib, subglabrous; bracteoles deltoid-ovate, 4–6.5 mm., subglabrous, pale and membranous, aristate with the excurrent pale midrib, the arista so little lengthening (to ± 2–3 mm.) in fruit that the spines of the burr still appear broad-based; ultimate divisions of cymes of a central fertile flower subtended on each side by a triad of 1 fertile and 2 modified flowers, or fertile flowers often absent except in the lower divisions. Outer 2 tepals lanceolate, 4–6 mm., rather broadly hyaline-margined, (4–)5–7-nerved in the central green portion with the nerves gradually reducing in length outwards, the midrib excurrent in a short mucro, the nerved portion ± densely long-pilose with barbellate multicellular hairs especially about the apex; inner 3 tepals shorter, 4–5 mm., oblong-lanceolate, 3–5-nerved, with a similar indumentum and also shortly mucronate. Modified flowers of a 2 small bracteoliform processes and 2 small straight spines. Filaments firm, compressed, 2.5–3.5 mm.; pseudostaminodes flabelliform, ± 1mm., dentate at the plane apex, with no dorsal scale. Ovary ovoid-subglobose. ± 1.5 ..

above; style slender, 2–2.5 mm. Capsule ovoid, ± 3 mm., slightly truncate but scarcely hardened at the apex. Seed ovoid, 2.75 mm., brown, almost smooth. Fig. 14/6, 7.

KENYA. Northern Frontier Province: 62 km. N. of turnoff to Merti from main Wajir–Isiolo Road, Dec. 1971, *Bally & Smith* 14688!; Tana River District: Tana River bank downstream from Garissa, Feb. 1956, *Greenway* 8864! & Garissa–Thika road, crossing of Tula Drift, June 1974, *R.B. & A.J. Faden* 74/799!

DISTR. **K** 1, 7; S. Somalia (Afmadu)

HAB. In *Lawsonia* thicket, in *Cordia, Ficus, Acacia* woodland, in riparian scrub and in semi-ruderal roadside vegetation, on grey sandy alluvium and grey, sticky clay; 30–300 m.

SYN. [*C. polycephala* sensu Chiov. in Fl. Somala 2: 378 (1932), *non* Bak.]
C. paniculata Hauman in B.J.B.B. 18: 109 (1946), *nom illegit.* sine descr. lat.; E.P.A.: 65 (1953)

NOTE. Apparently a very local species, which may be common where found.

10. **C. erinacea** *Schinz* in E.J. 21: 189 (1895) & P.O.A. C: 173 (1895); F.T.A. 6(1): 45 (1909); U.K.W.F.: 135 (1974). Type: "Ost-Afrika", *Fischer* 70 (Z, holo.!)

Erect, rather bushy, slightly to considerably branched annual herb, 0.1–1 m. tall; the lowest 1–2 internodes subterete, otherwise the stem and branches clearly angular-striate, pale with the striae darker, moderately or rather thinly furnished with thick-based bristly hairs or finally glabrescent; nodes ± swollen, but the stem and branches not shrunken above them when dry. Leaves oblong-ovate to broadly elliptic, 2–7 × 1.5–4 cm., acute to rather blunt at the apex, at the base cuneate into the 0.7–2 cm. petiole, moderately furnished with strigulose hairs or subglabrous with the hairs only on the lower surface of the primary venation and along the thickened, frequently reddish coloured margins. Inflorescences terminal on the stem and branches, from bud right through to fruit of solitary or paired globose heads of agglomerate cymes (very rarely a second lower pair on the same peduncle, not seen from the Flora region), in flower 0.75–1 cm. in diameter, in fruit enlarging and forming a burr up to 2.5–3 cm. with strongly accrescent very rigid spines 5–7 mm. long; peduncle 0.5–2 cm., thinly to densely white pilose; bracts broadly deltoid-ovate, 2.5–4 mm., whitish-membranous, thinly pilose, shortly and finely aristate with the excurrent mucro; bracteoles similar, usually smaller and more pilose, often asymmetrical with the midrib not central; ultimate divisions of cymes of a central fertile flower subtended on each side by a triad of 1 fertile and 2 modified flowers. Tepals all oblong, 3–5-nerved with the inner pair of lateral nerves occasionally branched, sometimes converging on the midrib above but not confluent with it, the outer nerves when present much shorter, the midrib excurrent in a short, obscure mucro; outer tepal longest and broadest, very concave, 3.5–4 mm., densely long-pilose dorsally except on the rather broad hyaline border; inner 4 tepals progressively shorter, glabrous or almost so, the inner 2 frequently slightly denticulate at the apex. Modified flowers of 2 narrow bracteoliform processes and 2–4 simple yellow or brownish spines. Filaments delicate ± 1.5 mm.; pseudostaminodes broadly cuneate-obovate, 0.5–0.75 mm., denticulate or excavate at the plane or slightly inflexed apex, frequently reddish coloured, with no dorsal scale. Ovary small, broadly ovoid, ± 1 mm., style very short, 0.3–0.5 mm. Capsule compressed-subrotund, broader than high, ± 3–3.5 mm. wide, firm-walled. Seed compressed-rotund, ± 2.5–3 mm., brown, almost smooth.

UGANDA. Karamoja District: Kangole, Aug. 1954, *J. Wilson* 136!

KENYA. Northern Frontier Province: Lokichar, July 1930, *Liebenberg* 286!; Nairobi/Machakos District: Athi River Station, Aug. 1947, *Bogdan* 1106!; Masai District: 54 km. Nairobi–Magadi road, Feb. 1969, *Greenway & Napper* 13560!

TANZANIA. Masai District: Ardai Plains, June 1944, *Greenway* 7000!; Mbulu District: Sambala, May 1929, *B.D. Burtt* 2150!; Pare District: Same, May 1928, *Haarer* 1239!

DISTR. **U** 1; **K** 1, 4, 6, 7; **T** 1–3, 5; Ethiopia

HAB. An apparently uncommon species, yet recorded from a diverse range of habitats – grassland (sometimes under shrubs), abandoned cultivation or in seasonal swampland, with soils ranging from black clay ("cotton soil") to saline alluvium; 500–1700 m.

SYN. *Sericocomopsis erinacea* (Schinz) Peter in F.D.O.-A. 2: 231 (1932)

11. **C. lanceolata** *Schinz* in E.J. 21: 188 (1895) & P.O.A. C: 173(1895); C.C. Townsend in Publ. Cairo Univ. Herb. 7–8: 76(1977). Type: "Ostafrika", *Fischer* 256 (Z, lecto.!)

Perennial herb, much branched from the base upwards, erect and bushy to prostrate and sprawling, 15–90 cm.; stem and branches strongly striate-sulcate, terete or some of the upper internodes tetragonous, whitish to green or the striae brown, subglabrous to pilose with softer and ± appressed to spreading and substrigose multicellular hairs; nodes slightly swollen, the stem and branches sometimes slightly shrunken above them. Leaves firm in texture, oblong to narrowly lanceolate-elliptic, 1.2–5.5 × 0.3–1.5 cm., acute to obtuse at the apex with a frequently deciduous firm horn-like mucro up to 2 mm. long, at the base cuneate to abruptly rounded or subauriculate, sessile or with a petiole up to 5 mm. long, both surfaces±softly appressed-pilose or appressed (more rarely ± spreading) strigose when young, finally thinly hairy or glabrescent or with strigose hairs persisting particularly along the sometimes undulate-crispate margins and along the lower surface of the midrib and few primary nerves. Inflorescences terminal on the stem and branches, solitary, rounded-capitate (formed of condensed cymes), the heads 1.5–2 cm. in diameter in flower and scarcely larger in fruit, commonly sessile and subtended by a pair of leaves, or on a pilose peduncle up to 1(–3) cm. long; bracts broadly ovate, broader than long, ± 3 mm., pale, glabrous, shortly aristate with the excurrent midrib; bracteoles broadly deltoid-ovate, 4–6mm., similar but more aristate, glabrous or thinly pilose above the base; ultimate divisions of cymes of a central fertile flower subtended on each side by a triad of a central fertile and 2 lateral modified flowers, or sterile flowers solitary or absent. Outer 2 tepals narrowly oblong-lanceolate, 6.5–9mm., firm, with broad pale opaque margins, with a narrow ($\frac{1}{3}$ of the width of the tepal or less) green vitta which ceases below the apex, 3(–5)-nerved in the vitta with the lateral nerves and midrib sometimes forked, lateral nerves evanescent above, the midrib not or very slightly excurrent in an obscure mucro; inner 3 tepals similar but slightly shorter, progressively blunter and the innermost subcucullate at the apex, the midrib not excurrent in a mucro but occasionally protruding as a small dorsal cusp near the apex, green vitta in all 3 paler with the nerves more clearly defined; indument of all 5 tepals variable, commonly evanescent, from slightly pilose about the base or furnished with long subappressed white hairs in the basal half (especially centrally along the pale margins) to rather densely floccose throughout. Modified flowers commonly lanate-hairy centrally, of a few narrowly bracteoliform processes and/or 2–8 very unequal stramineous to purplish spines which become accrescent and thickened at the base in fruit, the longest (3.5–)7–10 mm. in length. Filaments firm, 3–4 mm., alternately longer and shorter; pseudostaminodes considerably fused to the filaments, 1.5–2 mm., oblong, lacerate-dentate at the plane apex with no dorsal scale. Ovary squat, broadly pyriform (onion-shaped), tapering into the style, ± 1 mm.; style slender, 3.5–6 mm., pale and firm; Capsule ovoid, ± 3 mm., firm in the upper half and rupturing when ripe in the hyaline lower half. Seed subglobose, ± 2.75 mm., brownish, smooth. Fig. 14/1–5.

TANZANIA. Mwanza District: Massanza I., Nyambiti, Mar. 1953, *Tanner* 1281!; Lushoto District: 8 km. SE. of Mkomazi, May 1953, *Drummond & Hemsley* 2327!; Mbulu District: Singida–Babati road near Mt. Hanang, Mar. 1965, *Richards* 19958!

DISTR. T1–5, 7; Zimbabwe, Botswana, Namibia and South Africa

HAB. Habitats range from open or *Acacia* grassland to roadsides, saline flats, swamps or even in standing water, mostly on heavy soils from pale buff sandy clay loam to black cotton soil; 450–1690 m.

FIG. 14. *CYATHULA LANCEOLATA*—**1**, flowering branch, × $\frac{2}{3}$; **2**, tepals, × 5.5; **3**, part of androecium, × 10; **4**, gynoecium, × 5.5; **5**, sterile flower, × 10. *C. CORIACEA*—**6**, fruiting head, × 2; **7**, sterile flower with bracteole, × 10. 1–5, from *Richards* 19958; 6, 7, from *Bally* 9486. Drawn by Mary Millar Watt.

SYN. *C. crispa* Schinz in E.J. 21: 188 (1895). Type: South Africa, Transvaal, Makapansberge, *Rehmann* 5420 (K, iso.!)
C. merkeri Gilg in E.J. 36: 207 (1905); F.T.A. 6(1): 47 (1909). Type: Tanzania, Masai steppe, *Merker* (B, holo. †)
Pandiaka deserti N.E. Br. in K.B. 1909: 134 (1909). Type: Botswana, near Chukutsa salt-pan, *Lugard* 221 (K, holo.!)
P. lanceolata (Schinz) C.B. Cl. in F.T.A. 6(1): 68 (1909); F.D.O.-A. 241 (1932)
Cyathula hereroensis Schinz in Viert. Nat. Ges. Zürich 66: 222 (1921). Type: Namibia, Etosha Pan, *Dinter* 731 (Z, iso.!)
C. merkeri Gilg var. *strigosa* Suesseng. in. F.R. 44: 45 (1938). Type: Tanzania, between Singida and Minyugi [Manyugi], *B.D. Burtt* 1376 (K, holo.!)
C. deserti (N.E. Br.) Suesseng. in F.R. 44: 46 (1938)
C. strigosa Suesseng. in Mitt. Bot. Staats., München 1: 5 (1950). Type: Tanzania, Shinyanga, *Koritschoner* 2111 (K, holo.!)
Kyphocarpa ("*Cyphocarpa*" sphalm.) *kuhlweiniana* Peter in F.R. Beih. 40(2), Descr.: 26 (1932), Type: Tanzania, Lushoto District, Lake Manka, *Peter* 41059 (B, holo.!)
K. kuhlweiniana Peter var. *melanacantha* Peter in F.R. Beih. 40(2), Descr.: 27 (1932). Type: Tanzania, S. Pare, Buiko, *Peter* 10434 (B, holo.!)
Sericocomopsis lanceolata (Schinz) Peter in F.D.O.-A. 2: 229 (1932)
S. lanceolata (Schinz) Peter var. *merkeri* (Schinz) Peter in F.D.O.-A. 2: 229 (1932)
Pandiaka wildii Suesseng. in Mitt. Bot. Staats., München 1: 63 (1950). Type: Zimbabwe, Lower Sabi, *Wild* 2313 (SRGH, holo.!, K, iso.!)

11. **ALLMANIOPSIS**

Suesseng., Mitt. Bot. Staats., München 1: 2 (1950)

Low bushy perennial herb with entire alternate leaves. Inflorescence solitary, sessile, axillary, globular or very shortly ovoid, congested-cymose, bracteate; partial inflorescence basically of ultimate triads of a central fertile and 2 lateral sterile modified flowers formed of bracteoliform processes, but apparently variable and sometimes one modified flower absent or replaced by a second fertile flower; fruiting head possibly falling entire. Perianth-segments 5, decreasing in size inwards. Stamens 5, very short, the filaments delicate and monadelphous at the base, alternating with very short irregularly triangular pseudostaminodes; anthers minute, bilocular. Ovary with a single pendulous ovule, depressed-ovoid, glabrous; style short, bifid above with 2 linear stigmatal branches. Fruit a thin-walled capsule, rupturing irregularly. Seed compressed-ovoid; endosperm copious.

A monotypic genus.

A. fruticulosa *Suesseng.*, Mitt. Bot. Staats., München 1: 2 (1950); Cavaco in Mém. Mus. Nat. Hist. Nat. Paris, sér. B, 13: 92 (1962). Type: Kenya, Northern Frontier Province, Ijara, *Bally* 2179 (K, holo.!)

Low bushy perennial 12–25 cm. tall, with a very stout tough rootstock and numerous woody stems from about the base, glabrous in the older parts with a greyish brown striate cortex, young shoots ± densely whitish tomentellous. Leaves narrowly obovate to broadly spathulate, thick with cartilaginous margins, 12–30 × 5–12 mm., greyish green, secondary venation not apparent, subacute to subtruncate-apiculate at the apex, distinctly pale-mucronate with the excurrent midrib, attenuate and indistinctly petiolate below. Inflorescence whitish, 5–10 × 5–8 mm. Bracts and bracteoles similar or the former narrower, ± 2.5–3 mm., with a broadly ovate, glabrous or dorsally pilose membranous basal portion which is ± erose above and shortly tapering to truncate or incised on each side of the long outwardly curving or abruptly reflexed whitish arista (frequently longer than the base) formed by the excurrent midrib. Sterile flowers formed exclusively of long-aristate bracteoliform processes. Tepals of fertile flowers ovate, margins erose above

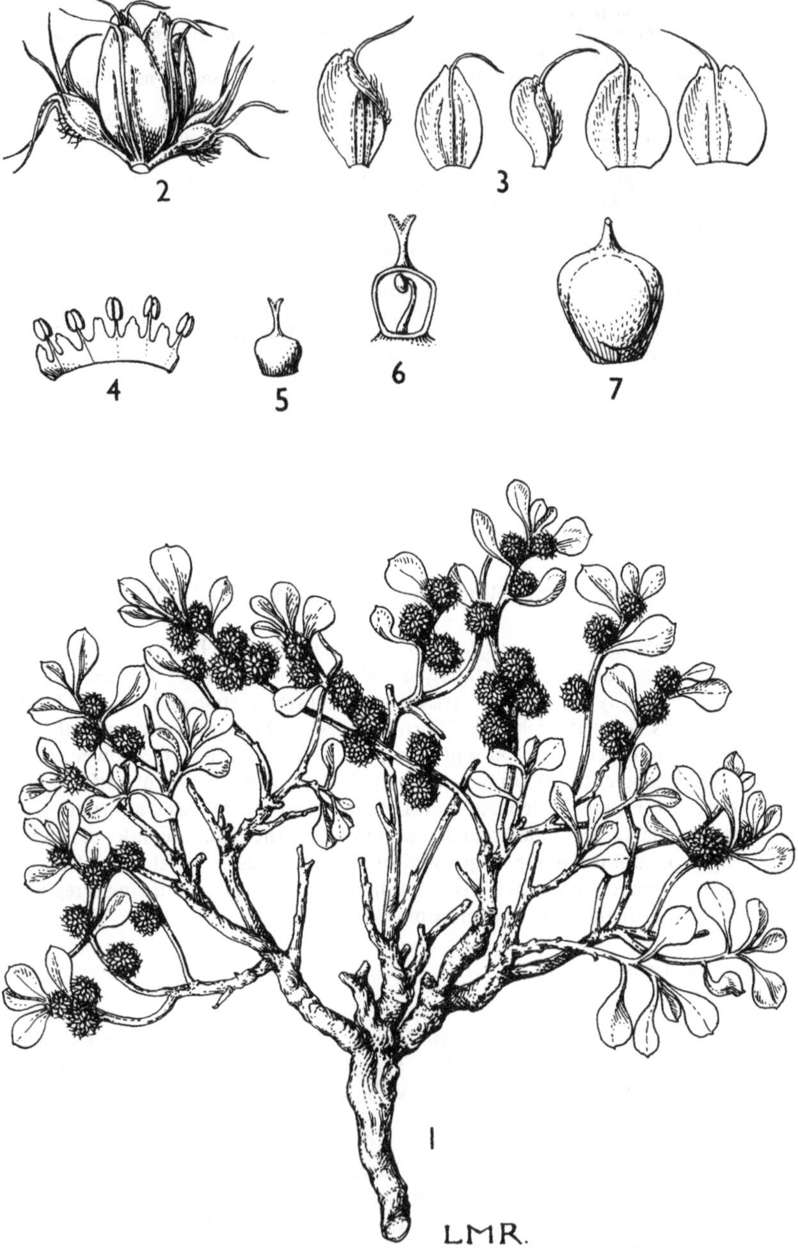

FIG. 15. *ALLMANIOPSIS FRUTICULOSA*—**1**, flowering plant, × $\frac{2}{3}$; **2**, triad of one fertile and two sterile flowers, ×6; **3**, tepals of fertile flower, × 6; **4**, portion of androecium, × 6; **5**, gynoecium, ×6; **6**, same, opened to show posture of ovule, ×8; **7**, fruit, × 10. 1–4, from *Bally & Radcliffe-Smith* 14493; 5, 6, from *Gillett* 16528; 7, from *Bally & Melville* 15248. Drawn by Lura Ripley.

and frequently incised-bilobulate at the apex, broadly hyaline-margined with a strongly 3-nerved green basal vitta, much reducing in size from the outermost (± 3.5 mm. including the long sharply reflexed arista) to the innermost (± 2 mm. including the short apical mucro), ± white pilose dorsally along the green vitta. Filaments very short and delicate, ± 0.5 mm.; pseudostaminodes ± equalling the filaments, broadly irregularly deltoid, entire or with an apical notch. Style ± 0.75 mm., the linear stigmas divergent, about equalling the column. Capsule ovoid-pyriform, 1.25–1.5 mm., firmer above the middle. Seeds compressed-ovoid, ± 1–1.25 mm., chestnut to blackish, shining; testa faintly reticulate. Fig. 15.

KENYA. Northern Frontier Province: 80 km. SW. of Wajir, *Bally & Smith* 14493!; Tana River District: Bilbil, 7.5 km. NW. of SKT 14, *Gillett* 16528! & 48 km. W. of Galole, Mar. 1964, *Makin* in *E.A.H.* 13065!

DISTR. **K**1, 7; not known elsewhere

HAB. Open *Acacia-Commiphora* bushland, on alkaline red sandy loam and pinkish-grey sandy soil; 110–200 m.

12. **PUPALIA**

A. Juss. in Ann. Mus. Hist. Nat. Paris 2: 132 (1803); C.C. Townsend in K.B. 34: 131–142 (1979), *nom. conserv.*

Pupal Adanson, Fam. Pl. 2: 268, 596 (1763)

Annual or perennial herbs or subshrubs with entire opposite leaves. Inflorescence a spiciform bracteate thyrse terminal on the stem and branches, each bract subtending a single ⚥ flower on each side of which is set a bracteolate modified flower consisting of a number of sharply hooked spines, or, more commonly, each bract containing a ⚥ flower subtended by 2 or more such triads each contained within a large bracteole. Spines of modified flowers at first very small, rapidly accrescent, finally disposed in 3 or occasionally more stalked clusters of 5–20 spines in 1–3 ranks, the clusters stellately or occasionally dendroidly set on a common peduncle, subequalling to much exceeding the perianth and serving as a means of distributing the fruit; bracts persistent, finally ± deflexed; entire partial inflorescences falling intact in fruit. Perianth-segments 5. Stamens 5; filaments delicate to rather solid, fused at the extreme base on to a fleshy lobed disk-like cup into which the base of the ovary is narrowed; pseudostaminodes absent; anthers bilocular. Ovary with a single pendulous ovule; style slender; stigma capitate. Fruit a thin-walled capsule, irregularly ruptured below the firm apex by the developing seed. Seed oblong-ovoid or ovoid, slightly compressed; endosperm copious.

4 species in the tropics (extending to the subtropics) of the Old World from W. Africa to Malaysia and the Philippines.

Bracts each subtending a single fertile flower, i.e. one triad of 1 fertile and 2 modified sterile flowers; tepals 2.75–3.5 mm. . . . . 1. *P. micrantha*
Most or all of the bracts subtending more than one fertile flower; tepals only exceptionally as short as 3.5 mm.:
  Tepals narrowly oblong-lanceolate, (6–)7–8 mm,; style long and slender, (2.75–)3–3.5 mm. . . . . 2. *P. grandiflora*
  Tepals oblong-ovate, (3–)4–5(–6) mm.; style shorter, very rarely attaining 3 mm., and mostly less than 2 mm. 3. *P. lappacea*

1. **P. micrantha** *Hauman* in B.J.B.B. 18: 109 (1946) & in F.C.B. 2: 61

(1951); C.C. Townsend in K.B. 34: 133 (1979). Type: Zaire, Shaba, Lofoi R., *Quarré* 5649 (BR, holo.!)

Annual herb ± 0.5–1.5 m. tall, bushy or straggling with many divaricate branches from the base upwards (small plants simple); stem and branches slender, terete, striate, thinly or in parts more densely furnished with whitish multicellular hairs, the older basal internodes finally glabrescent. Leaves lanceolate-ovate to elliptic, acuminate, those of the stem and main branches 5.5–15 × 2.9–5.4 cm. including the 0.5–1.8 cm. petiole, dark green and thinly rather long-pilose on the upper surface, paler and more densely and shortly pubescent beneath, subtruncate to cuneate at the base; leaves of upper part of stem and branches rapidly reducing in size. Inflorescences considerably elongating as the flowers open and finally up to ± 25 cm. long including the (to ± 8 cm.) peduncle, solitary or paired in the leaf-axils, up to 4–5 together at the ends of the stem and branches, simple or sometimes branched, axis thinly to densely pilose. Bracts lanceolate-ovate, 1–1.5 mm., membranous-margined, persistent, ± thinly pilose, each subtending a single triad of 1 fertile and 2 modified bracteolate flowers. Bracteoles broadly deltoid-ovate, ± 2 mm., abruptly shortly acuminate with a short sharp mucro formed by the excurrent midrib, broadly membranous-margined below, moderately pilose. Tepals oblong-ovate, 2.75–3.5 mm., broadly white-margined, 3-nerved in the green centre with the nerves confluent above to form a short sharp mucro; outer 2 tepals slightly longer and broader, lanate chiefly about the base, the inner 3 lanate over most of the dorsal surface. Style short, ± 0.5 mm. Sterile flowers of 3 branches each bearing ± 5–8 hooked setae to ± 2 mm. long, not forming a dense burr nor concealing the fertile flower. Fruit an ovoid, somewhat compressed capsule ± 1.75–2 mm. long, rupturing irregularly at the thin-walled base. Seed round, somewhat compressed, ± 1.5–1.75 mm., black, almost smooth, shining. Fig. 16/8, 9.

TANZANIA. Near Shinyanga, *Bax* 41!; Mpanda District: Kasekela stream valley, Apr. 1964, *Pirozynski* 691!; Ufipa/Mpanda District: Rukwa Rift Valley, Mar. 1947, *Pielou* 138!
DISTR. T1, 4; Ivory Coast, Nigeria, Ethiopia, Zaire, Zambia, Malawi, Mozambique, Madagascar
HAB. Open deciduous woodland, near streams or in seasonal swamps; 760–970 m.

SYN. *Pupalia psilotrichoides* Suesseng. in Mitt. Bot. Staats., München 1: 64 (1950). Type: Mozambique, Namagoa, *Faulkner* 14 (K, holo.!)
Cyathula prostrata (L.) Blume var. *grandiflora* Suesseng. in Mitt. Bot. Staats., München 1: 77 (1951). Type: Tanzania, Rukwa Rift Valley, *Pielou* 138 (K, lecto.!)
Sericorema humbertiana Cavaco in Bull. Mus. Nat. Hist. Nat. Paris, sér. 2, 24: 574 (1952). Type: Madagascar, *Humbert* 12471 (P, lecto.!)
[*Digera alternifolia* sensu F.W.T.A., ed. 2, 1: 148 (1954), *non* (L.) Aschers.]

2. **P. grandiflora** *Peter* in F. R. Beih. 40(2), Descr.: 22 (1932); F.P.N.A. 1: 134 (1948); Hauman in F.C.B. 2: 61 (1951); C.C. Townsend in K.B. 34: 134 (1979). Type: Tanzania, S. Pare Mts., Tona–Mbaga, *Peter* 8617 (B, lecto.!)

Perennial herb, often rather woody at the base, scandent, trailing or more rarely erect, 1–2(–4) m., much branched; stem and branches weak, terete, striate, thinly to moderately furnished with whitish multicellular hairs, the older basal internodes finally glabrescent. Leaves lanceolate to broadly ovate, acuminate, those of the stem and branches 3.2–14 × 2.2–6 cm. including the 1–2.5 cm. petiole, dark green and thinly rather long-pilose on the upper surface, paler and more densely and shortly pubescent beneath (rarely tomentose on the midrib and principal veins), rounded to cuneate at the base; upper leaves of stem and branches rapidly reducing in size. Inflorescences considerably elongating as the flowers open and finally up to 35(–48) cm. long including the (up to 9 cm.) peduncle, solitary and terminal on the stem and branches, axis moderately spreading pilose or densely tomentose. Bracts lanceolate, 3–4 mm., darkly membranous-margined, persistent, moderately pilose, each subtending a partial inflorescence of 3–7 fertile flowers, most of which are set between 2 modified sterile flowers, but the central solitary. Bracteoles of triads of 1 fertile and 2 sterile flowers broadly deltoid-ovate, ± 4 mm., abruptly shortly acuminate with a sharp yellowish to dark mucro formed by the excurrent midrib, broadly membranous-margined below, moderately densely pilose dorsally. Bracteoles of sterile flowers ovate-lanceolate, ± 4 mm., membranous with a green midrib which is excurrent in a distinct brownish arista, thinly to moderately pilose. Tepals (6–)7–8 mm., 3-nerved in the green centre with the nerves confluent above to form a short sharp mucro, narrowly oblong-lanceolate; outer 2 tepals slightly longer, ± uniformly long-pilose, narrowly membranous-margined, the inner 3 more broadly pale-margined (not conspicuously so since the margins are incurved), more densely long-pilose. Style long and slender, (2.75–)3–3.5 mm. Sterile flowers dendroidly branched with several divaricate branches each ending in (6–)9–15(–20) hooked setae up to ± 6 mm. long, usually brownish but occasionally yellow, forming a very dense globose burr ± 1.5–2.2 cm. in diameter, concealing the fertile flowers. Fruit an oblong-ovoid capsule 2–2.25 mm. long, rupturing irregularly at the thin-walled base. Seed ovoid, ±2 mm., black, almost smooth, shining. Fig. 16/5–7.

UGANDA. Kigezi District: Mitano Gorge, Feb. 1947, *Purseglove* 2326!; Masaka District: 2 km. E. of Kayugi, May 1972, *Lye* 6934!; Mengo District: Mukono, Oct. 1913, *Dummer* 350!
KENYA. Northern Frontier Province: Mt. Kulal, Oct. 1947, *Bally* 5587!; Nairobi District: Karura Forest, Oct. 1967, *Mwangangi & Abdalla* 233!; Masai District: Ngong Forest, Jan. 1934, *Napier* in *C.M.* 5870!
TANZANIA. Masai District: Ngorongoro Forest, Nov. 1932, *Geilinger* 3691!; Pare District: Mamba, May 1927, *Haarer* 411!; Lushoto District: Mkusu valley, Mkuzi-Kifungilo, Apr. 1953, *Drummond & Hemsley* 2231!
DISTR. U2, 4; **K**1, 3, 4, 6; **T**2, 3, 6; Zaire, Rwanda, Sudan, Ethiopia, N. Yemen
HAB. Mostly at forest edges, along rides and in clearings, also in open woodland, in bush and along rivers – apparently always in more or less undisturbed habitats; 1150–2000 m.

SYN. *P. lappacea* (L.) A. Juss. var. *grandiflora* (Peter) Suesseng. in Mitt. Bot. Staats., München 1: 7 (1950)

3. **P. lappacea** (L.) *A. Juss.*, Ann. Mus. Hist. Nat. Paris 2: 132 (1803); P.O.A.C : 173 (1895); F.T.A. 6(1): 47 (1909); F.D.O.-A. 2: 225 (1932); F.P.N.A. 1: 133 (1948); F.P.S. 1: 122 (1950); Hauman in F.C.B. 2: 60 (1951); Cavaco in Mém. Mus. Nat. Hist. Nat. Paris, sér. B, 13: 90 (1962); F.F.N.R.: 45 (1962); F.P.U.: 102 (1962); U.K.W.F.: 136 (1974); C.C. Townsend in K.B. 34: 135 (1979) Type: Sri Lanka, *Hermann Herbarium* Vol. 1, p. 2, larger piece (BM, lecto.!)

Annual or perennial herb, ± erect and ± 0.3–0.9 m. tall, or prostrate and sprawling, or subscandent and scrambling to as much as 2.5 m.; stem generally much branched and swollen at the nodes, branches opposite, divaricate or ascending, slender, obtusely 4-angled to almost terete, thinly pilose to densely tomentose. Leaves variable in shape and size, from narrowly ovate-elliptic to oblong or circular, 2–10(–14) × 1–5(–7) cm., acuminate to obtuse-apiculate or retuse at the summit, shortly or more longly cuneate at the base, narrowed to a petiole 2–2.5(–3.5) cm. long; indumentum of lamina varying from sericeous or tomentose to subglabrous with a few hairs running vertically along the lower surface of the primary venation, rarely quite glabrous, commonly moderately pilose with the hairs along the nerves divergent. Inflorescences at first ± dense, elongating to as much as 0.5 m. in fruit with the lower flowers becoming increasingly remote; axis subglabrous to tomentose; peduncle ± 1–10 cm.; bracts lanceolate, 1.5–2.5 mm., persistent, ± deflexed after the fall of the fruit, subglabrous or pilose, sharply mucronate with the percurrent midrib; partial inflorescences mostly of 1 solitary ⚥ flower subtended on each side by a triad of 1 ⚥ and 2 modified flowers; bracteoles of each triad broadly subcordate-ovate, (2.75–)3–5(–6) mm., abruptly narrowed to the stramineous to dark arista formed by

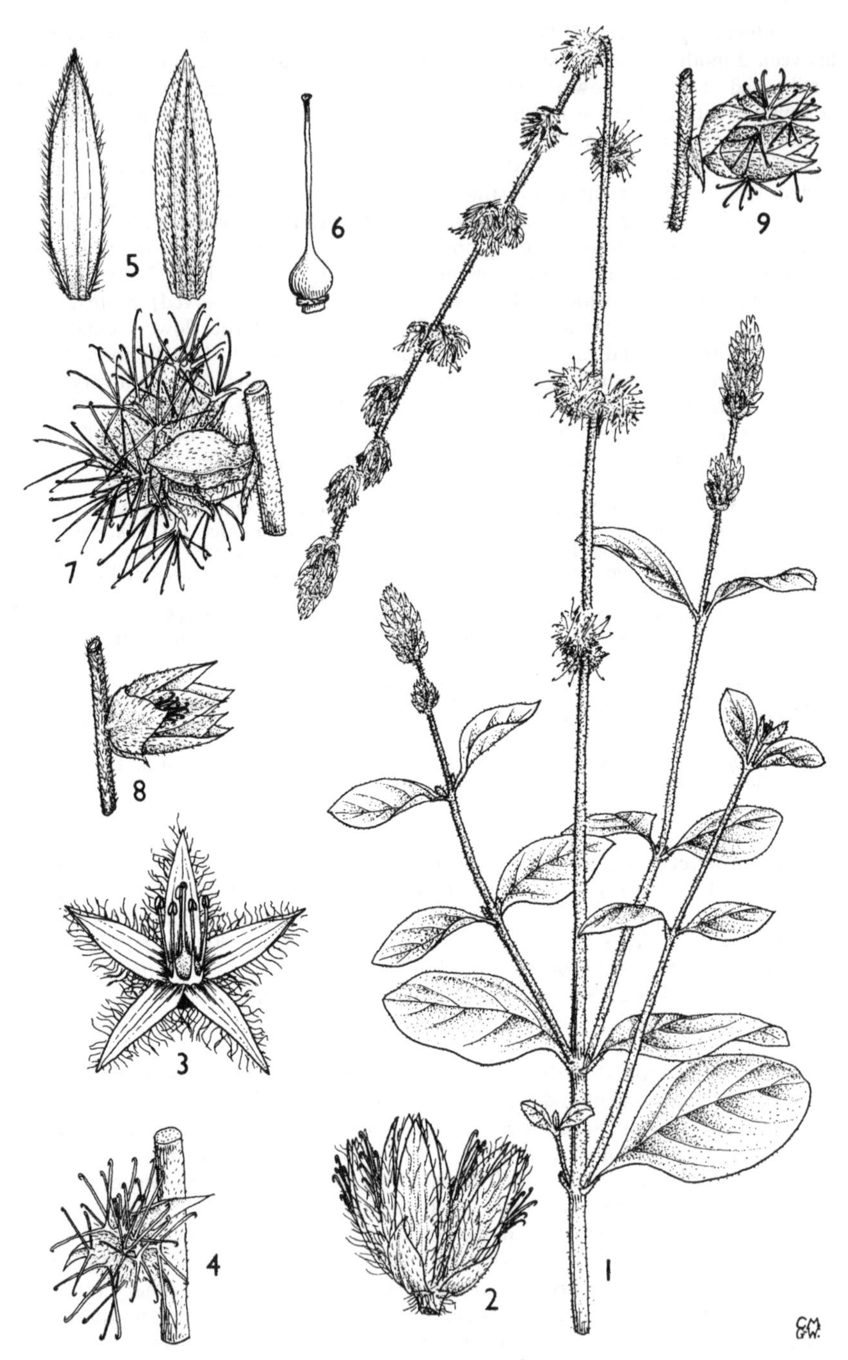

FIG. 16. *PUPALIA LAPPACEA* var. *VELUTINA*—**1**, flowering branch, × $\frac{2}{3}$; **2**, triad of one fertile and two sterile flowers, × 4; **3**, flower opened up, × 4. **4**, fruiting partial inflorescence, × 2. *P. GRANDIFLORA*—**5**, outer tepals, × 4; **6**, gynoecium, × 4; **7**, partial inflorescence in young fruit, × 2. *P. MICRANTHA*—**8**, partial inflorescence, × 4; **9**, same in fruit, × 4. 1, 4, from *Thomas* 3416; 2, 3, from *Gillett* 12888; 5, 6, from *Mshana* 221; 7, from *Drummond & Hemsley* 1393; 8, from *Pirozynski* 691; 9, from *Fanshawe* 9139. Drawn by Christine Grey-Wilson.

the excurrent midrib, membranous with a pale margin, thinly to very densely hairy; bracteoles of sterile flowers ovate-lanceolate, usually more shortly and less densely pilose. Tepals oblong-ovate to lanceolate-ovate, ± quickly narrowed to a rather obtuse-mucronate apex to gradually narrowed and acute-aristate, the outer 2 (3–)4–5(–6) mm., subglabrous to ± tomentose dorsally, 3(–5)-nerved, the midrib and 2 inner nerves confluent just below the apex and excurrent in the mucro or short arista, inner 3 slightly shorter and more densely pilose. Branches of sterile flowers 3, each terminating in (3–)5–18(–20+) setae in 1–3 ranks; setae subglabrous to ± villous in the lower half, yellowish to purple or red, (1.5–)3–7 mm., the partial inflorescence falling intact to form a burr ± 8–18 mm. in diameter. Style short to rather slender, (0.5–)0.9–2(–3) mm. Capsule ovoid, 2–2.5 mm. Seed oblong-ovoid with a prominent radicle, 2 mm. long, dark brown, shining; testa at first faintly reticulate but finally smooth or punctulate.

DISTR. (of species as a whole) **U**1–3; **K**1–7; **T**1–8; widespread in the tropics of the Old World; throughout tropical Africa N. to Egypt, also in South Africa and Madagascar, Arabia and Asia from India eastwards to Malaya, the Malayan Is. (Java, Celebes, etc.), the Philippines and New Guinea; introduced in Australia, etc.

SYN. *Achyranthes lappacea* L., Sp. Pl.: 204 (1753)
A. atropurpurea Lam., Encycl. Méth. 1: 546 (1785). Type: specimen grown in Royal Gardens at Paris, *Herb. Lamarck* (P-LA, lecto.!, IDC microfiche 546.4!)
Pupalia atropurpurea (Lam.) Moq. in DC., Prodr. 13(2): 331 (1849); P.O.A.C: 173 (1895); F.T.A. 6(1): 48 (1909); E.P.A.: 66 (1953)

NOTE. Var. *lappacea*, a thinly pilose plant with a slender inflorescence given a "spiky" appearance by the tapering sharply acute tepals, is found in S. India, Sri Lanka, South Africa and Mozambique. It has not been found in the region of the Flora, though a few specimens (e.g. *Harris, Poćs & Tadros* 4328 from Pongwe, Tanga District, Tanzania) diverge towards it.

KEY TO INFRASPECIFIC VARIANTS

Outer 2 tepals glabrous on the back or sparingly pilose on the lower half, with tufts of hairs at the basal angles; vegetative parts glabrous or very sparingly shortly hairy .... b. var. **glabrescens**
Outer 2 tepals pilose to lanate; vegetative parts thinly pilose to tomentose:
Outer tepals lanceolate-ovate, tapering gradually to the mucronate tip; branches of sterile flowers terminated by (8–)12–18(–21) usually red setae up to 5 mm.; leaves silvery-sericeous beneath especially when young .... a. var. **argyrophylla**
Outer tepals oblong-ovate, rather abruptly narrowed to the mucronate tip; branches of sterile flowers terminated by (4–)5–8(–13) usually straw-coloured setae up to 7 mm,; leaves thinly pilose to tomentose, but rarely silvery-sericeous .... c. var. **velutina**

a. var. **argyrophylla** *C.C. Townsend* in K.B. 34: 140 (1979). Type: Tanzania, Pangani District, Bushiri, *Faulkner* 592 (K, holo.!)

Leaves constantly ovate, silvery-sericeous on the lower surface especially when young, the midrib and primary venation very conspicuous. Inflorescence-axis densely furnished with patent or deflexed white hairs. Outer 2 tepals lanceolate-ovate, 4–5 mm., tapering gradually to the acutely mucronate apex, furnished with long white hairs, 3-nerved. Branches of the sterile flowers terminated with (8–)12–18(–21) slender uncinate setae up to 5 mm. long, which are (where colour notes are available) red at least when young.

KENYA. Kwale District: 3.2 km. from Kwale on Tanga road, Feb. 1968, *Magogo & Glover* 125!; Kilifi District: Sokoke, Aug. 1936, *Moggridge* 157!; Lamu District: Boni Forest, Mararani, 20 Sept. 1961, *Gillespie* 383!

TANZANIA. Tanga District: Sawa, May 1965, *Faulkner* 3522!; Uzaramo District: Dar es Salaam, *Stuhlmann* 7563!; Rufiji District: Mafia I., Mar. 1933, *Wallace* 780! in part.; Zanzibar I., Chaani, Jan. 1929, *Greenway* 1199!; Pemba I., Mvumoni, Aug. 1929, *Vaughan* 551!

DISTR. **K**7; **T**3, 6; **P**; **Z**; not known elsewhere

HAB. Coastal bushland and grassland, forest edges, weed of cultivation; 0–530 m.

b. var. **glabrescens** *C.C. Townsend* in K.B. 34: 139 (1979). Type: Zanzibar I., Makunduchi, *Faulkner* 2332 (K, holo.!)

Vegetative parts of plant glabrous or very sparingly shortly hairy, dull green or blackish when dry. Stem usually red at least about the nodes. Inflorescence-axis glabrous or sparingly pilose. Outer 2 tepals rather broadly oblong-ovate, 4.5–5 mm., with lanate tufts of hairs at the basal angles, dorsal surface glabrous or sparingly pilose in the lower half, occasionally 4–5-nerved. Branches of the sterile flowers terminated with 5–11 uncinate setae up to 3 mm. long. Burrs 8–9 mm. in diameter.

KENYA. Kwale District: Jadini, June 1956, *Irwin* 254!; Lamu District: Kiungamwina [Kuingamini] I., July 1961, *Gillespie* 55!

TANZANIA. Uzaramo District: Oyster Bay, June 1968, *Batty* 142!; Rufiji District: Chole I., Sept. 1937, *Greenway* 5304!; Kilwa District: Nyuni I., Oct. 1974, *Frazier* 2099!; Zanzibar I., Chukwani, June 1960, *Faulkner* 2800!; Pemba I., Mvumoni, Aug. 1929, *Vaughan* 544!

DISTR. **K**7; **T**6, 8; **Z**; **P**; Mozambique

HAB. Sandy ground by sea shore and a little inland, coral cliffs, arable land and waste places; 0–6 m.

SYN. [*P. atropurpurea* sensu Moq. in DC., Prodr. 13(2): 331 (1849) quoad pl. Pembae]

c. var. **velutina** *(Moq.) Hook. f.*, Fl. Brit. Ind. 4: 724 (1885). Type: Burma, banks of R. Irrawaddy, *Wallich Herb.* 6935A (K, iso.!)

Plant very variable in leaf shape and indumentum, from thinly pilose to densely villous or tomentose, but rarely silvery-sericeous. Inflorescence-axis thinly hairy to densely furnished with spreading or deflexed hairs. Outer tepals oblong-ovate, (3.5–)4–5(–6) mm., rather quickly narrowed to the mucronate apex, ± densely pilose to lanate even in forms with thinly pilose foliage, 3-nerved. Branches of the sterile flowers terminated by (4–)5–8(–13) usually stramineous uncinate setae up to 7 mm. long. Fig. 16/1–4.

UGANDA. Karamoja District: Moroto, May 1940, *A.S. Thomas* 3416!; Bunyoro District: Bulisa, Jan. 1941, *Purseglove* 1100!; Teso District: Serere, Mar. 1932, *Chandler* 618!

KENYA. Northern Frontier Province: 11 km. S. of Kangetat, Lokori, May 1970, *Mathew* 6230!; Machakos District: Kibwezi, Dec. 1921, *Dummer* 5048!; Teita District: Tsavo East National Park, E. of Voi, Jan. 1970, *Ryvarden* 5025!

TANZANIA. Masai District: E. Serengeti, Masandari, July 1962, *Newbould* 6166!; Lushoto District: 3 km. WSW. of Mkomazi on Moshi road, Mar. 1975, *Hooper & Townsend* 1031!; Iringa District: Ruaha National Park across Mwagusi R. from Mbagi, Feb. 1966, *Richards* 21299!

DISTR. **U**1–3; **K**1–7; **T**1–8; through the range of the species, of which it is much the commonest variety.

HAB. In a great variety of habitats – forest edges and clearings, grassland, bushland abandoned and current arable ground, gravelly and sandy lake shores, roadsides, dry thickets, among lava rock etc., chiefly on sandy soils or clay with sand admixed; 10–1900 m.

SYN. *Desmochaeta flavescens* DC., Cat. Hort. Monsp.: 102 (1813), *nom. illegit.*
Achyranthes mollis Thonn. in Schumach. & Thonn. Beskr. Guin. Pl.: 137 (1827). Type: Ghana, *Thonning* 163 (C, holo., IDC microfiche of Herb. Isert & Thonning 2.6!)
Pupalia velutina Moq. in DC., Prodr. 13(2): 332 (1849)
P. mollis (Thonn.) Moq. in DC., Prodr. 13(2): 333 (1849)
P. distantiflora A. Rich., Tent. Fl. Abyss. 2: 217 (1850). Type: Ethiopia, Choho, *Quartin Dillon* (P, holo.!)
P. tomentosa Peter in F.R. Beih. 40(2), Descr.: 23 (1932). Type: Tanzania, Lushoto District, NW. of Buiko, *Peter* 40919 (B, holo.!)
P. brachystachys Peter in F.D.O.-A. 2: 227 (1938), *nomen.*
P. lappacea (L.) A. Juss. var. *tomentosa* (Peter) Suesseng. in Mitt. Bot. Staats., München 1: 31 (1954)

NOTE. *Pupalia affinis* Engl. in Hans Meyer, O.-Afr. Gletscherfahrten: 369 (1890), *nomen* (cf. F.T.A. 6(1): 49), was revived by Peter in F.D.O.-A. 2: 227 (1983), but remains undescribed.

13. DASYSPHAERA

Gilg in E. & P. Pf., Nachtr. 2–4, 1: 153 (1897)

Perennials or small shrubs with opposite or more rarely alternate leaves and branches; indumentum of jointed, almost smooth hairs. Leaves entire. Inflorescence a narrow bracteate thyrse with compact sessile partial inflorescences; partial inflorescences basically of ultimate bracteolate triads of 1 fertile and 2 bibracteolate sterile flowers; fertile flowers 3–6 in each partial inflorescence, commonly with 2 triads set on each side of an ebracteolate fertile flower. Fertile flowers ⚥ with 5 perianth-segments which are all similar in form. Stamens 5; filaments filiform throughout or expanded towards the base, free or very shortly connate at the base; pseudostaminodes absent; anthers bilocular. Ovary with a single pendulous ovule, glabrous, obpyriform, firm and rounded above, delicate below; style filiform, stigma capitate. Sterile flowers of few narrow bracteoliform processes and numerous smooth or pilose bristles in several branched clusters which lengthen as the fertile flowers mature. Fruit an indehiscent capsule rupturing in the lower delicate area, the entire partial inflorescence probably falling together in fruit. Endosperm copious.

4 species in E. and NE. Africa.

D. tomentosa Lopr. in E.J. 27: 53 (1899)*; Schinz in Viert. Nat. Ges. Zürich 56: 254 (1911). Type: Tanzania, Moshi District, Lake Chala, *Volkens* 1800 (Z, iso.!)

Perennial suffrutescent herb, ± 0.6–1 m. tall, with numerous simple or usually branched stems from about the base; cortex of older parts light brown to greyish; young branches terete, whitish-tomentose, the tomentum becoming less dense and yellowish with age. Leaves subrotund to ± reniform, the main stem leaves commonly broader than long (smaller leaves about as broad as long), lamina 0.9–2.5 × 1.1–3.5 cm. in length and breadth, ± densely appressed tomentose, the lower surface whiter especially along the principal veins and midrib, obtuse to apiculate at the apex, shortly (3–7 cm.) petiolate. Inflorescence cylindrical, 6–14 × 2.5–3.5 cm., rather dense or the partial inflorescences becoming remote below, on a 1–2.5 cm. peduncle. Partial inflorescences dense, sessile. Bracts deltoid-lanceolate, 4–8 mm., membranous with green vitta, distinctly excurrent in a brownish awn, densely tomentose. Bracteoles of ultimate triads similar to the bracts or somewhat narrower; bracteoles of sterile flowers narrowly lanceolate, 6–9 mm., more longly aristate, much less pilose especially about the base. Tepals 5, narrowly lanceolate, the outer not infrequently somewhat falcate at the tip, 8–10.5 mm., narrowly hyaline-margined in the basal half, with 3 prominent green to brownish nerves, midrib excurrent in a sharp brownish arista, dorsal surface of nerves densely long-pilose with almost smooth hairs; inner tepals gradually somewhat shorter and narrower. Stamens 3–6 mm., filaments filiform throughout, free almost or quite to the base. Style 3–4 mm. Sterile flowers of 2–3 lanceolate-subulate bracteoliform processes and numerous unequal, smooth, slender, spiny bristles. Partial inflorescences in fruit probably falling entire as a blackish-brown bristly ball ± 2.5–3 cm. in diameter. Capsule ovoid, 3–3.5 mm., smooth, greenish brown. Seed 2–2.5 mm., brown, shining; testa feebly reticulate. Fig. 17.

*First mentioned, without description, by Gilg in E. & P. Pf., Nachtr. 2–4, 1: 153 (1897), with the name attributed to Volkens.

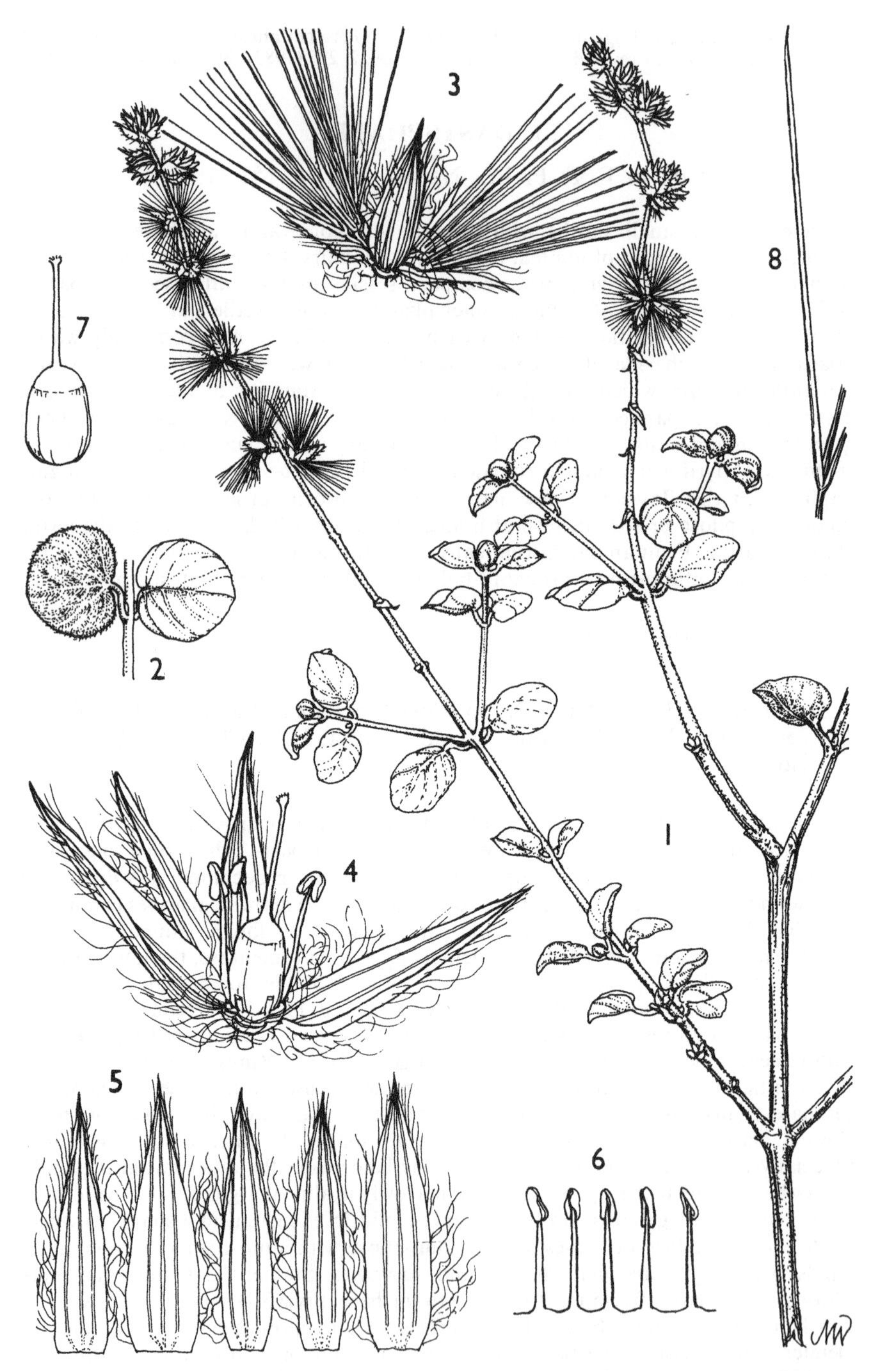

FIG. 17. *DASYSPHAERA TOMENTOSA*—**1**, flowering branch, × $\frac{2}{3}$; **2**, pair of leaves, × $\frac{2}{3}$; **3**, triad of one fertile and two sterile flowers, ×2.5; **4**, fertile flower with one tepal and two stamens removed, × 6; **5**, tepals, × 5.5; **6**, part of androecium, × 6; **7**, gynoecium, × 6; **8**, bristle of sterile flower, × 5. 1–7, from *Greenway* 4467; 8, from *Napier* 951. Drawn by Mary Millar Watt.

KENYA. Teita District: Maratate, May 1931, *Napier* 951! & Murka Camp, July 1966, *V.C. Gilbert* C79! & foot of Maktao Hill, Aug. 1969, *Bally* 13413!
TANZANIA. Masai District: Olmoti, Mar. 1977, *Peterson* 372!; Moshi District: Lake Chala, Jan. 1936, *Greenway* 4467!; Pare District: Same, June 1958, *Mahinde* 234!
DISTR. **K**7; **T**2, 3; not known elsewhere
HAB. Open grassland and *Acacia* bushland; 900–1520 m.

SYN. *Marcellia tomentosa* (Lopr.) C.B. Cl. in F.T.A. 6(1): 52 (1909); F.D.O.-A. 2: 228 (1932)

14. VOLKENSINIA

Schinz in Viert. Nat. Ges. Zürich 57: 535 (1912)

Kentrosphaera Gilg in E. & P. Pf. Nachtr. 2–4, 1: 153 (1897), *non* Borzi, Studi Algologici 1: 87 (1883)

Bushy perennial or small shrub with opposite leaves and branches; indumentum of jointed, minutely barbellate hairs. Leaves entire. Inflorescence a narrow bracteate thyrse with compact pedunculate partial inflorescences; partial inflorescences basically of ultimate bracteolate triads of 1 fertile and 2 bracteolate sterile flowers, 2 triads normally set on each side of an ebracteolate fertile flower, but fertile flowers sometimes scattered only. Fertile flowers ⚥ with 5 perianth-segments which are all similar in form. Stamens 5; filaments filiform, alternating with truncate pseudostaminodes which are adnate to the filaments for most of their length to form a basal membrane; anthers bilocular. Ovary with a single pendulous ovule, glabrous, turbinate, rounded above with a circumferential free "skirt" about the middle, appearing as if furnished with a scutellate cap; style filiform; stigma capitate. Sterile flowers of narrow bracteole-like processes and smooth bristles which greatly lengthen as the fertile flowers mature. Fruit a thin-walled indehiscent capsule, the entire partial inflorescence apparently falling in fruit. Endosperm copious.

A monotypic genus.

V. prostrata *(Gilg) Schinz* in Viert. Nat. Ges. Zürich 57: 535 (1912). Type: Tanzania, Kilimanjaro, Kifaru [Nashornhügel] *Volkens* 472 (K, iso.!)

Much-branched perennial herb or subshrub with a tough fibrous rootstock and brittle branches, 0.15–1.37 m. (rarely trailing to ± 3 m.), the old wood with a greyish to whitish, glabrescent, frequently cracking cortex; young twigs and branches striate, ± densely appressed canescent. Leaves ovate or broadly ovate to narrowly or oblong-lanceolate, lamina 2–6.5(–11) × 1.4–3.5(–6), moderately to densely appressed canescent on both surfaces, obtuse to subacute at the apex, attenuate or abruptly narrowed at the base into a distinct 0.7–2.5(–4) cm. petiole. Inflorescence lax and narrow, 6–15 × 2–5 cm., on a long (2–9.5 cm.) peduncle. Partial inflorescences dense; peduncles up to ± 2 cm., slender. Bracts deltoid-ovate, 3–5 mm., membranous with a narrow green very shortly excurrent midrib, ± white-pilose with appressed hairs dorsally. Bracteoles of ultimate triads similar to the bracts or rather smaller; bracteoles of sterile flowers lanceolate-subulate, more longly aristate, ± 3–4 mm., also dorsally pilose. Flowers bright carmine-red. Tepals 5, narrowly lanceolate, 6–8 mm., ± densely furnished over the dorsal surface with appressed, jointed, minutely barbellate white hairs, 3-nerved with the midrib very shortly excurrent at the narrow but subcucullate apex. Stamens 4–5 mm. Style 1.75–4 mm. Partial inflorescences in fruit falling entire as a golden or ± pink-tinged bristly ball ± 2–4 cm. in diameter. Capsule obovoid with a keel around the apex, ± 2 mm., brown or blackish. Seed ± 1.75 mm., brown, smooth; testa feebly reticulate. Fig. 18.

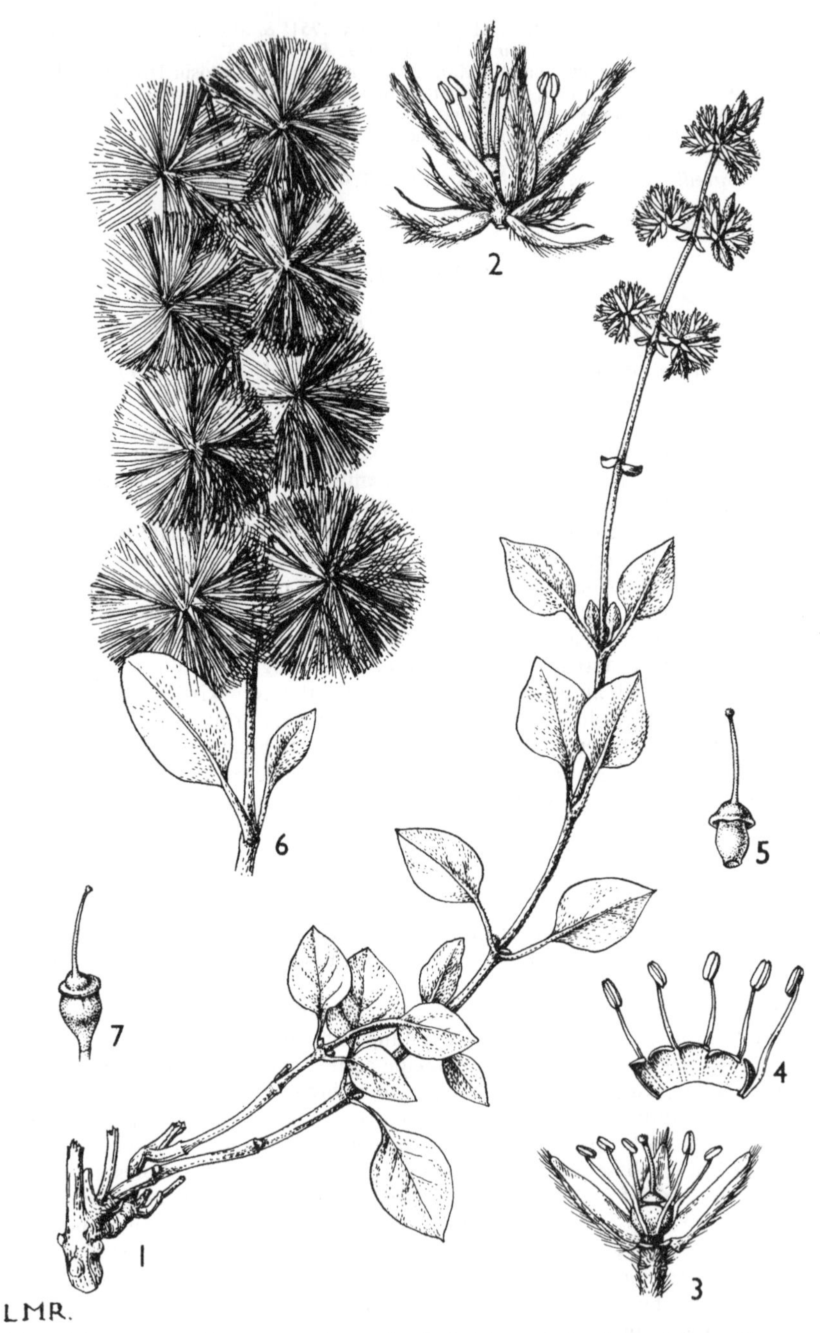

FIG. 18. *VOLKENSINIA PROSTRATA*—**1**, flowering branch, × $\frac{2}{3}$; **2**, triad of one fertile and two sterile flowers, ×2; **3**, flower with two tepals removed, ×2; **4**, part of androecium, inner surface, ×4; **5**, ovary, ×4; **6**, fruiting branch, × $\frac{2}{3}$; **7**, fruit, ×6. 1-5, from *Mathew* 6446; 6, from *Newbould* 6966; 7, from *Makin* 239. Drawn by Lura Ripley.

KENYA. Northern Frontier Province: Archer's Post, Nov. 1965, *Makin* 239!; Masai District: edge of Lake Amboseli, Mar. 1977, *Hooper & Townsend* 1318!; Tana River District: Garissa, Jan. 1972, *Gillett* 19505!
TANZANIA. Masai District: Ngaserai, June 1965, *Leippert* 5832!; Moshi District: Arusha Chini-Kahe, Sept. 1964, *Beesley* 6!; Lushoto District: Mkomazi, July 1955, *Semsei* 2156!
DISTR. K1-4, 6, 7; T2, 3, ?5, 6; SE. Sudan, S. Ethiopia
HAB. Semi-desert, *Acacia-Commiphora* bushland, dry *Cenchrus-Cynodon* grassland, flood plains, crevices and soil pockets among lava outcrops, edges of pools and lakes, etc. - on dry or well-drained soils from grey-brown loam and black cotton soil to pure sand or powdery volcanic dust; 180–1360 m.

SYN. *Kentrosphaera prostrata* Gilg in E. & P. Pf., Nachtr. 2-4, 1: 153 (1897)
Marcellia prostrata (Gilg) C.B. Cl. in F.T.A. 6(1): 51 (1909); F.D.O.-A. 2: 228 (1932)
M. leptacantha Peter in F.R. Beih. 40(2), Descr.: 26 (1938). Type: Tanzania, Lushoto District, Mkomazi to Lake Manga, *Peter* 10835 (B, holo.!)
Volkensinia grandiflora Susseng. in Mitt. Bot. Staats., München 1: 70 (1955). Type: Kenya, Northen Frontier Province, Sololo, *Gillett* 13677 (K, holo.!, EA, iso.)
V. prostrata (Gilg) Schinz forma *lanceolata* Suesseng. in Mitt. Bot. Staats., München 1: 77 (1955). Type: Kenya, Northern Frontier Province, Archer's Post, *Edwards* in *A.D.* 2989 (K, lecto.!)
Dasysphaera prostrata (Gilg) Cavaco in Mém. Mus. Nat.Hist. Nat. Paris, sér. B, 13: 96 (1962); U.K.W.F.: 136 (1974)

15. AERVA

Forssk., Fl. Aegypt.-Arab.: 170 (1775), *nom. conserv.*

Ouret Adans., Fam. Pl. 2: 268, 586 (1763)

Perennial herbs (sometimes flowering in the first year), prostrate to erect or scandent. Leaves and branches opposite or alternate, leaves entire. Flowers ⚥ or dioecious, sometimes probably polygamous, bibracteolate, in axillary and terminal sessile or pedunculate bracteate spikes, one flower in the axil of each bract. Perianth-segments 5, oval- or lanceolate-oblong, membranous-margined with a thin to wider green centre, the perianth deciduous with the fruit but bracts and bracteoles persistent. Stamens 5, shortly monadelphous at the base, alternating with subulate or rarely narrowly oblong and truncate or emarginate pseudostaminodes; anthers bilocular. Ovary with a single pendulous ovule; style very short to slender and distinct; stigmas 2, short to long and filiform (sometimes solitary and capitate, flowers then probably functionally ♂). Capsule thin-walled, bursting irregularly. Seed compressed-reniform, firm, black.

About 10 species in the tropics, chiefly centred on Africa.

Outer 2 tepals 2–3 mm., the midrib ceasing well below the apex; individual spikes up to ± 10 cm. in length .... 1. *A javanica*
Outer 2 tepals 0.75–1.5(–2) mm., the midrib excurrent in a short mucro; individual spikes mostly up to 4(–6) cm. in length:
  Spikes sessile, axillary, solitary or in clusters, forming an elongate inflorescence which is leafy to the apex of the stem and branches .... 2. *A. lanata*
  Spikes sessile or shortly pedunculate, towards the ends of the stem and branches forming leafless terminal racemes, the entire inflorescence thus paniculate .... 3. *A. leucura*

1. **A. javanica** *(Burm. f.) Schultes*, Syst. Veg., ed. 15, 5: 565 (1819); P.O.A. C: 173 (1895); F.P.S. 1: 113 (1950); Cavaco in Mém. Mus. Nat. Hist. Nat.

Paris, sér. B, 13: 100 (1962). Type: *Herb. Burmann* (G, holo.!), alleged to have come from Java, but the species is unknown there

Perennial dioecious herb, frequently woody and suffruticose or growing in erect clumps, 0.3–1.5 m., branched from about the base with simple stems or the stems with long ascending branches. Stem and branches terete, striate, ± densely whitish or yellowish tomentose or pannose, when dense the indumentum often appearing tufted. Leaves alternate, very variable in size and form, from narrowly linear to suborbicular, ± densely whitish or yellowish tomentose but usually more thinly so and greener on the upper surface, margins plane or ± involute (when strongly so the leaves frequently ± falcate-recurved), sessile or with a short and indistinct petiole or the latter rarely to ± 2 cm. in robust plants. Spikes sessile, cylindrical, dense and stout (up to ± 10×1 cm.), to slender and interrupted with lateral globose clusters of flowers and with some spikes apparently pedunculate by branch reduction; ♂ plants always with more slender spikes (but plants with slender spikes may not always be ♂); upper part of stem and branches leafless, the upper spikes thus forming terminal panicles; bracts broadly deltoid-ovate, 0.75–2.25 mm., hyaline, acute or obtuse with the obscure midrib ceasing below the apex, densely lanate throughout or only about the base or apex, persistent; bracteoles similar, also persistent. Female flowers with outer 2 tepals 2–3 mm., oblong-obovate to obovate-spathulate, lanate, acute to obtuse or apiculate at the tip, the yellowish midrib ceasing well below the apex; inner 3 slightly shorter, elliptic-oblong, ± densely lanate, acute, with a narrow green vitta along the midrib, which extends for about two-thirds the length of each tepal; style slender, distinct, with the 2 filiform flexuose stigmas at least equalling it in length; filaments reduced, anthers absent. Male flowers smaller, the outer tepals 1.5–2.25 mm., ovate; filaments delicate, the anthers about equalling the perianth; ovary small, style very short, stigmas rudimentary. Capsule 1–1.5 mm., rotund, compressed. Seed round, slightly compressed, 0.9–1.25 mm., brown or black, shining and smooth or very faintly reticulate. Fig. 19/4, 5.

UGANDA. Karamoja District: Lokitonyala, Sept. 1956, *J. Wilson* 274!
KENYA. Northern Frontier Province: El Wak, May 1952, *Gillett* 13342!; Turkana District: Lorengipe, Oct. 1963, *Bogdan* 5636! (♂); Masai District: Ol Lorgosailie Plains, July 1947, *Bally* 5162!
TANZANIA. Masai District: base of Longido Mt., Mar. 1970, *Richards* 25687!; Pare District: Nyumba ya Mungu, Aug. 1968, *Batty* 271!; Mpwapwa District: between Gulwe and Msalgali, Mar. 1947, *van Rensberg* 567!
DISTR. **U**1; **K**1–4, 6, 7; **T**2, 3, 5–7; widespread in the drier parts of the tropics and subtropics of the Old World from Morocco south to Cameroun, across the drier regions of Africa to Egypt, the Sudan and Somalia south to Madagascar, in Asia from Palestine and Arabia to Burma, India and Sri Lanka: adventive in Australia and elsewhere
HAB. Very varied, deciduous woodland and bushland to grassland and disturbed places, rocky volcanic gorges and dry hill with lava outcrops, soil heavy or more commonly light and sandy; 0–1520 m.

SYN. *Celosia lanata* L., Sp. Pl.: 205 (1753). Type: Sri Lanka, *Hermann Herbarium* Vol. 4, p. 52, L.H. specimen (BM, lecto.!)
Iresine persica Burm. f., Fl. Indica: 212 (sphalm. 312), t. 65/1 (1768). Type: Iran, *Herb. Burmann* (G, holo.!)
I. javanica Burm. f., Fl. Indica: 212 (sphalm. 312), t. 65/2 (1768)
Illecebrum javanicum (Burm. f.) Murr., Syst. Veg., ed. 13: 206 (1774)
Aerva tomentosa Forssk., Fl. Aegypt.-Arab.: cxxii, 170 (1775); F.T.A. 6(1): 37 (1909); F.D.O.-A. 2: 217 (1932). Type: Egypt, *Forsskal* (C, lecto., IDC microfiche 3.19!)
Achyranthes javanica (Burm.f.) Pers., Syn. 1:259 (1805)
A. persica (Burm.f.) Merr. in Philipp. Journ. Sci. 19: 348 (1921); E.P.A.: 69 (1953); U.K.W.F.: 136 (1974)

NOTE. Forms of *A. javanica* with slender inflorescences have occasionally been misidentified as *A. artemisioides* Vierh. & Schwartz in E. and NE. Africa. Apart from a more bushily branching habit, the latter is easily recognised by the ± capitate stigma which is sessile or subsessile on the apex of the ovary; the broad, obtuse or notched staminodes; and the tepals being all ± similar. It is a tropical Arabian species.

2. **A. lanata** *(L.) Schultes*, Syst. Veg., ed. 15, 5: 564 (1819); P.O.A. C: 173 (1895); F.T.A. 6 (1): 39 (1909); F.D.O.-A. 2: 218 (1932); F.P.N.A. 1: 134 (1948); F.P.S. 1: 115 (1950); Hauman in F.C.B. 2: 57 (1951); E.P.A.: 68 (1953); Cavaco in Mém. Mus. Nat. Hist. Nat. Paris. sér. B, 13: 103 (1962); F.P.U.: 104 (1962); U.K.W.F.: 136 (1974). Type: *Herb. Linnaeus* 290.6 (LINN, lecto.!)

Perennial herb, sometimes flowering in the first year, frequently woody and suffrutescent below, prostrate to decumbent or erect (occasionally scandent), stiff or weak and straggling, (0.1–)0.3–2 m., branched from the base and often also above (upper branches short to long and slender even on the same plant), but stem and main basal branches often not further branched. Stem and branches terete, striate, ± densely lanate with whitish ± shaggy hairs, more rarely tomentose or canescent. Leaves alternate, round to lanceolate- or ovate-elliptic, shortly or more longly cuneate at the base, rounded and apiculate to acute at the apex, commonly densely lanate or canescent on the lower surface and more thinly so above, sometimes subglabrous on the upper surface, rarely glabrous throughout or thickly lanuginose, those of the main stem 10–50 × 5–35 mm., those of the branches and upper part of the stem reducing and often becoming very small; petioles from 2 cm. down to almost none. Spikes sessile, solitary or usually in axillary clusters on the main stems or long to very short axillary branches, 0.4–1.5(–2) × 0.3–0.4 cm., divergent, cylindrical, silky, white to creamy, forming a long inflorescence which is leafy to the ultimate spikes; bracts deltoid-ovate to oblong-ovate, 0.75–1 mm., membranous with a short arista formed by the excurrent midrib, pilose, persistent; bracteoles similar or slightly smaller, also persistent. Tepals ± densely lanate dorsally. Flowers ♀ or ⚥ (probably also sometimes functionally ♂). Hermaphrodite flowers: outer 2 tepals hyaline, oval-oblong, ± abruptly contracted at the tip to a distinct mucro formed by the excurrent nerve, 0.75–1.75 mm. without the mucro; inner 3 slightly shorter and narrower, acute with a broad central green vitta along the midrib, which extends for about three-quarters of the length of each tepal and is often ± furnished with a thickened border; style short, stigmas very short and patent or divergent; anthers perfect; probable male flowers similar but stigmas reduced, subcapitate or very short, scarcely papillose. Female flowers: tepals sometimes similar to those of the ⚥ flowers but commonly longer and narrower, tapering above, the outer up to ± 2.25 mm., style slightly longer, stigmas distinctly longer and often equalling the style, linear, divergent or suberect; anthers absent, filaments reduced. Capsule rotund, compressed ± 1 mm. Seed reniform, ± 0.6–0.8 mm., black, shining, the testa almost smooth in the centre, faintly reticulate around the margin.

UGANDA. Karamoja District: eastern Matheniko, near Turkana Scarp, Mar. 1959, *J. Wilson* 701! Kigezi District: Kachwekano Farm, July 1949, *Purseglove* 3013!; Mengo District: Bukasa sandpits, Aug. 1952, *Lind* 88!

KENYA. Northern Frontier Province: S. Turkana, Loriu plateau, June 1970, *Mathew* 6570!; Kitui District: Mutha Plains, Jan. 1942, *Bally* 1632!; Mombasa, Port Tudor, *MacNaughton* 96!

TANZANIA. Ngara District: Bugufi, Rusumo, Dec. 1960, *Tanner* 5383!; Arusha District: Meru National Park, Lake Longil, E. side, *Greenway & Kanuri* 11973!; Bagamoyo District: Mandera Primary School, Sept. 1969, *Mwakalasi* DSM. 951!; Zanzibar I., Chukwani, Feb. 1963, *Faulkner* 3169!; Pemba I., Ras Kigomasha, Dec. 1930, *Greenway* 2689!

DISTR. **U**1–4; **K**1–7; **T**1–8; **Z**; **P**; widespread in the drier parts of the tropics and subtropics of the Old World – in Africa from Sierra Leone across to Egypt, S. to South Africa (rare) and Madagascar, also in Seychelles, Chagos Archipelago, etc; in Asia from Arabia E. to Malaysia, Malayan Is., the Philippines and New Guinea

HAB. Very varied, from cultivated and disturbed ground to woodland, bushland and grassland, at swamp and forest edges, open lava screes and lava boulder-strewn hillsides, coastal sands, etc; 0–2030 m.

SYN. *Achyranthes lanata* L., Sp. Pl.: 204 (1753)
Illecebrum lanatum (L.) L., Mant. Pl. Alt.: 344 (1771)
Achyranthes villosa Forssk., Fl. Aegypt.-Arab.: 48 (1775). Type: *Forsskål Herb.* 203 (C, holo., IDC microfiche 2.9!)

Aerva lanata (L.) Schultes var. *oblongata* Aschers. in Schweinf., Beitr. Fl. Aeth.: 174 (1867); F.T.A. 6(1): 40 (1909); F.D.O.-A.: 220 (1932); F.P.S. 1: 115 (1950); E.P.A.: 68 (1953). Type: Ethiopia, Dschadscha, *Schimper* (holo. not located)
A. lanata (L.) Schultes var. *intermedia* Suesseng. in B.J.B.B. 15: 57 (1938). Type: Zaire, Shaba, Kitimbo, *Kassner* 2346 (BR, holo.!)
A. lanata (L.) Schultes var. *leucuroides* Suesseng. in B.J.B.B. 15: 57 (1938). Types: Zaire, 3 *Quarré* syntypes (BR)
A. incana Suesseng. in Mitt. Bot. Staats., München 1: 1 (1950). Type: Kenya, Kwale District, Tiwi [Tivi]. *Ritchie* in *Bally* 2702 (K, holo.!), *non* Mart. (1826)
A. sansibarica Suesseng. in K.B. 4: 475 (1950). Type: Zanzibar, Pwani, Mchangani, *Greenway* 1202 (K, holo.!, EA, iso.)
A. lanata (L.) Schultes var. *elegans* Suesseng. in Mitt. Bot. Staats., München 1: 66 (1950). Type: Kenya, Kilifi, *Jeffery* 346 (EA, holo.!)
A. lanata (L.) Schultes forma *grandifolia* Suesseng. in Mitt. Bot. Staats., München 1: 66 (1950). Type: Tanzania, Shinyanga, *Koritschoner* 2110 (EA, holo., K, iso.!)
A. lanata (L.) Schultes forma *microphylla* Suesseng. in Mitt. Bot. Staats., München 1: 66 (1950). Type: Tanzania, Arusha, *van Rensburg* 383 (EA, holo.!)
A. lanata (L.) Schultes var. *oblongata* forma *squarrosa* Suesseng. in Mitt. Bot. Staats. München 1: 66 (1950). Type: Kenya, Nairobi, *Milne* 12 (EA, holo., K, iso.!)
A. lanata (L.) Schultes var. *pseudojavanica* Suesseng. in Mitt. Bot. Staats., München 1: 67 (1950). Type: Tanzania, Masai District, Ardai Plains, *Couchman* 94 (EA, holo.)
A. lanata (L.) Schultes var. *suborbicularis* Suesseng. in Mitt. Bot. Staats., München 1: 67 (1950). Type: Pemba, *Greenway* 2689 (EA, holo., K, iso.!)
A. lanata (L.) Schultes var. *leucuroides* Suesseng. in Mitt. Bot. Staats., München 1: 70 (1951), *nom. illegit.* Type: Zimbabwe, Marandellas, *Wild* 3304 (K, iso.!)
A. lanata (L.) Schultes var. *rhombea* Suesseng. in Mitt. Bot. Staats., München 1: 334 (1953). Type: Kenya, Kilifi District, Malindi, *Tweedie* 973 (K, holo.!)

NOTE. The great variability of *A. lanata* and the difficulty of separating the variations even at infraspecific rank has been discussed elsewhere (Townsend in K.B. 29: 461–463 (1974).

3. **A. leucura** *Moq.* in DC., Prodr. 13(2): 302 (1849); F.T.A. 6(1): 39 (1909); F.D.O.-A. 2: 218 (1932); F.P.N.A. 1: 135 (1948); F.C.B. 2: 52 (1951); E.P.A.: 69 (1953); Type: South Africa, Griquatown, Leeuwenkiul valley *Burchell* 1892 (K, iso.!)

Perennial herb apparently sometimes flowering in the first year, frequently woody below, erect or low-spreading to prostrate or occasionally ± scrambling, 0.5–1.2(–1.5) m., with numerous stems arising from the base; stems simple or branched with short or long, ascending, slender branches. Stem and branches terete, striate, ± densely tomentose or lanate with whitish hairs. Leaves alternate, broadly elliptic to linear-oblanceolate, those of the main stem ± 1.6–10 × 0.3–3(–3.6) cm., shortly cuneate to attenuate and ± petiolate at the base with a petiole up to ± 1.5 cm. long, rather obtuse to very acute at the apex, moderately to thinly pilose on both surfaces or more sparsely so above, leaves of the branches and inflorescence reducing upwards. Inflorescence variable, always with a leafless terminal spike or lax or compact terminal panicle of simple or branched, sessile or pedunculate spikes, the upper leaf-axils for a variable length of the stem with similar pedunculate panicles, or these reduced to a sessile, simple, sometimes lobed spike. Ultimate spikes white, dense or occasionally laxer and more elongate, 0.6–5 (–8) × 0.4–1.2 cm., divergent or ascending; bracts deltoid-ovate to oblong-ovate, 1.25–1.75 mm., hyaline with a short or longer arista formed by the excurrent midrib, pilose, persistent; bracteoles similar or slightly smaller, also persistent. Flowers ⚥, ♀ or functionally ♂. Tepals ± densely lanate dorsally, ± similar in all types of flower; outer 2 tepals hyaline, elliptic-oblong, (1.5–)2–2.5 mm., with a sharp mucro formed by the excurrent nerve; inner 3 tepals slightly shorter and narrower, acute with a central green vitta which extends for about three-quarters the length of each tepal and is often ± furnished with a thickened border. Female flowers with linear stigmas somewhat shorter than or subequalling the style, divergent or suberect; anthers absent, filaments reduced. Male flowers with the stigmas very reduced or absent and only the truncate apex of the style slightly papillose, occasionally with short and apparently receptive stigmas but

FIG. 19. *AERVA LEUCURA*—**1**, flowering branch, × $\frac{2}{3}$; **2**, functionally male flower, × 12; **3**, female flower, with tepals removed, × 12. *A. JAVANICA*—**4**, female flower, × 12; **5**, male flower, with tepals removed × 12. 1, 2, from *Richards* 11195; 3, from *Anderson* 661; 4, from *Magogo* 1416; 5, from *Bogdan* 5636. Drawn by Christine Grey-Wilson.

setting no seed; anthers large. Hermaphrodite flowers with stigmas usually (but not invariably) shorter than in ♀, and anthers often somewhat smaller than in ♂. Capsule and seeds as in *A. lanata*. Fig. 19/1–3.

UGANDA. Ankole District: Ruhengere, Nov. 1950, *Jarrett* 16!; Busoga District: 3km. N. of Nkondo gombolola and W. of Nkondo-Kigingi [Kigindi] Pier road, July 1953, *G.H. Wood* 819!; Mengo District; 16km. NW. of Nakasongola, Sept. 1955, *Langdale-Brown* 1496!

KENYA. Meru District: Meru Game Reserve, June 1963, *Mathenge* 36!; Kitui District: Kibwezi-Kitui road 11km. after the Athi R., Apr. 1969, *Napper & Kanuri* 2034!; Masai District: Narok, June 1976, *Sturrock* 1130!

TANZANIA. Musoma District: Grumeti R., W. of Musabi, Feb. 1968, *Greenway, Kanuri & Brown* 13170!; Mpanda District: Tumba, Mar. 1951, *Bullock* 3779!; Dodoma District: Itigi-Chunya road, ± 16 km. from Itigi, Nov. 1960, *Richards* 13528!

DISTR. U1–4; K4, 6; T1–8; tropical Africa from Zaire to Angola and Namibia also in Zambia, Zimbabwe, Malawi, Botswana and South Africa

HAB. In varying situations from disturbed ground to deciduous bushland, woodland and forest edges, rocky places, open streamsides, etc.; 600–1670 m.

SYN. *A. ambigua* Moq. in DC., Prodr. 13(2): 302 (1849). Type: South Africa, Vaal R., *Burke* (K, holo.!)

A. leucura Moq. var. *lanatoides* Suesseng. in Mitt. Bot. Staats., München 1: 67 (1950). Type: Tanzania, Mbulu District, Mt. Hanang, *Greenway* 7751 (K, isolecto.!)

NOTE. The sexual diversity of this species in Africa would appear from herbarium material to have a distinct geographical pattern. In Kenya and Uganda, hermaphrodite plants are apparently frequent, as they are in northern Tanzania (T1, 2 & 6). In southern and western Tanzania, however (as in the "Flora Zambesiaca" region), hermaphrodite flowers are rare and plants are either female (anthers absent) or functionally male (stigmatal surface much reduced, and even if apparently receptive no seed produced).

16. NOTHOSAERVA

Wight, Ic. Pl. Ind. Or. 6: 1 (1853)

Pseudanthus Wight, Ic. Pl. Ind. Or. 5(2): 3 (1852), *non* Spreng. (1827)

Annual herb with opposite or alternate branches and leaves, Leaves entire. Flowers small, solitary in the axils of scarious bracts, in dense sessile solitary or clustered spikes. Perianth-segments 3–4(–5), hyaline; perianth subtended by 2 very small bracteoles. Stamens 1 or 2; filaments filiform, intermediate teeth absent; anthers bilocular. Ovary with a single pendulous ovule; radicle ascending; style short; stigma capitate. Capsule delicate, irregularly rupturing. Seeds rounded, compressed; endosperm copious.

A monotypic genus.

N. brachiata *(L.) Wight,* Ic. Pl. Ind. Or. 6: 1 (1853); P.O.A.C: 173 (1895); F.P.S. 1: 119 (1950); E.P.A.: 70 (1953); Cavaco in Mém. Mus. Nat. Hist. Nat. Paris, sér. B, 13: 104 (1962); U.K.W.F.: 136 (1974). Type: India, *Herb. Linnaeus* 290.1 (LINN, holo.!)

Annual herb, (4–)10–45 cm., with many spreading branches from about the base upwards; stem and branches subterete, striate, glabrous or thinly hairy. Leaves narrowly to broadly elliptic, elliptic-oblong or ovate, entire, thinly hairy to glabrous, obtuse to subacute at the tip, blade of the lower main stem leaves ± 10–40(–50) × 6–20 mm., gradually or more abruptly narrowed to a petiole about half the length of the blade, upper and branch leaves becoming shorter and narrower. Flowers in dense 3–15 × 2–2.5 mm. spikes which are clustered in the leaf-axils of the stem and branches or on very short axillary shoots; spike sessile or the terminal spike on axillary shoots shortly (to ± 3 mm.) pedunculate; inflorescence-axis thinly to rather densely pilose; bracts

FIG. 20. *NOTHOSAERVA BRACHIATA*—**1**, plant, × $\frac{2}{3}$; **2**, flower, × 30; **3**, bract, × 60; **4**, flower with one perianth segment removed, × 30; **5**, longitudinal section of ovary, × 30; **6**, fruit, × 30; **7**, seed, × 30. All from *Drummond & Hemsley* 4067. Drawn by Pat Halliday.

hyaline, minutely erose, concave, acute or shortly acuminate, ± 0.5 mm., glabrous or very thinly hairy; bracteoles minute, hyaline. Perianth-segments broadly ovate, ± 1.25 mm., subacute to shortly acuminate, villous on the outer surface in the basal two-thirds with a thick greenish vitta and a central rib. Stamens longer than ovary and style. Capsule included, falling with the persistent perianth. Seed ±0.4 m., chestnut-brown, smooth and shining. Fig. 20.

KENYA. Northern Frontier Province: 11 km. S. of Kangetet, Lokori, May 1970, *Mathew* 6276!; Baringo District: 3 km, SW. of Lake Baringo, Sept. 1959, *Bogdan* 4890!; Tana River District: Bura, Mar. 1963, *Thairu* 78!

TANZANIA. Dodoma District: S. of Dodoma, Ruaha R., Sept. 1932, *Geilinger* 1767!;Bagamoyo District: 28 km. NNW. of Dar es Salaam, Aug. 1972, *Wingfield* 2029/ A; Iringa District. Great Ruaha R., July 1956, *Milne-Redhead & Taylor* 11237!

DISTR. **K**1, 3, 4, 7; **T**5, 6–8; tropical Africa from Senegal through Chad to Somalia (N.) and Ethiopia, S. to Zimbabwe and Angola; Mauritius; India from Punjab to Madras and Sri Lanka, Burma: doubtfully recorded from Borneo

HAB. Usually in sandy or clay flats, shores or seasonal watercourses, where liable to periodic flooding, on bare soil by ditches, etc.; 0–1050 m.

SYN. *Achyranthes brachiata* L., Mant. Pl.: 50 (1767)
Illecebrum brachiatum (L.) L., Mant. Pl. Alt.: 213 (1771)
Aerva brachiata (L.) Mart. in Nov. Act. Acad. Caes.-Leop. Carol., Nat. Curios. 13(1): 291 (1826); F.T.A. 6(1): 40 (1909); F.D.O.-A. 2: 221 (1932)
Pseudanthus brachiatus (L.) Wight, Ic. Pl. Ind. Or. 5(2): 3, t. 1776 (1852)

17. PSILOTRICHUM

Blume, Bijdr. Fl. Nederl. Ind.: 544 (1826)

Psilostachys Hochst. in Flora 27, Beil.: 6, t. 4 (1844)

Perennial herbs or subshrubs, prostrate to scandent, with entire, opposite or partly alternate leaves. Flowers ⚥ in axillary and terminal bracteate heads or spikes, solitary in the axil of each bract and bibracteolate; bracts persistent, finally spreading or deflexed; bracteoles falling with the fruit. Tepals 5, free, strongly to faintly nerved or ribbed (nerves 3 or more), the outer 2 tepals frequently finally ± indurate at the base, usually differing in form and indumentum from the inner 2 with the middle tepal intermediate. Stamens 5, shortly monadelphous at the base, without or rarely with alternating pseudostaminodes; anthers bilocular. Ovary with a single pendulous ovule; style slender but rather short; stigma capitate. Capsule thin-walled, bursting irregularly. Seed ovoid, brownish.

About 14 species in tropical Asia and Africa, chiefly the latter.

Leaves linear .... 10. *P. schimperi*
Leaves not linear:
Flowers in a large, open, compound panicle, the ultimate branchlets and the axes of the ultimate spikes very slender and capillary, the latter ± zigzag:
Flowers 3–3.5 mm., nodes and internodes without or with few long, bristly, multicellular hairs (Fig. 21/7); leaves never silvery-sericeous beneath .... 1. *P. gnaphalobryum*
Flowers smaller, 2–2.5 mm.; stem and branches with tufts of long, bristly multicellular hairs between the branches and leaves of at least the upper nodes, commonly also along the internodes (Fig. 21/4); at least the young leaves very silvery-sericeous beneath .... 2. *P. sericeum*
Flowers in opposite axillary and terminal spikes, rarely paniculate and then the branches and axes of the ultimate spikes neither capillary nor zigzag:

Tepals with many fine and obscure nerves (Fig. 22/6)... 3. *P. scleranthum*
Tepals with few to several very prominent nerves:
Outer tepals with broad, pale, smooth margins almost or quite as wide as the central nerved portion... 7. *P. elliotii*
Outer tepals not or very narrowly pale-margined:
Tepals 2.5–3.5 mm.; fruiting perianth long-stipitate (± 2 mm.) .... 8. *P. cyathuloides*
Tepals 5–7 mm.; fruiting perianth not long-stipitate:
Tall perennial herb, 0.9–1.5 m.; outer 2 tepals 7–9-nerved, furnished with short appressed hairs 4. *P. majus*
Low growing perennials up to 0.3 m.; outer 2 tepals 5-nerved, furnished with spreading hairs:
Leaves considerably pilose on both surfaces; tepals lanceolate, acute .... 9. *P. vollesenii*
Leaves glabrous or almost so, at least on the upper surface; tepals lanceolate, rather finely acuminate:
Dorsal nerves of tepals prominent, but not carinate with acute channels between; inner tepals with the margins glabrous or very thinly long-pilose.................... 5. *P. fallax*
Dorsal nerves of tepals very prominent, carinate with acute channels between; inner tepals with dense, long, ascending white hairs along the margins .... 6. *P. axilliflorum*

1. **P. gnaphalobryum** *(Hochst) Schinz* in Viert. Nat. Ges. Zürich 57: 550 (1912); F.P.S. 1: 120 (1950); E.P.A.: 71 (1953). Type: Yemen, Mt. Gesser, *Schimper* Iter unio 1837 No. 785 (K, isolecto.!)

Perennial herb, somewhat woody at the base, much-branched and bushy, ± 1 m. tall, or scrambling over shrubs and then reaching a greater height. Stems and branches terete, striate and glabrescent in the older parts, when younger subquadrate, sulcate-striate, and softly pilose especially about the nodes; nodes scarcely swollen. Main stem and branch leaves broadly lanceolate to ovate or subcordate-ovate, gradually tapering from near the base to the acute or acuminate apex, 3.5–6.5 × 1.2–3.7 cm., leaves rapidly reducing above and finally small and linear-lanceolate in the partial inflorescences, all moderately to densely furnished with fine appressed hairs. Inflorescence a large open compound panicle, of axillary opposite branches formed of simple or branched spikes each with a zigzag thinly pilose rhachis; peduncle and especially the branches of the partial inflorescences very slender, the latter capillary, thinly pilose with fine hairs; bracts ovate-lanceolate, ± 1 mm., hyaline with the greenish or brownish midrib excurrent in a very short mucro, furnished with fine soft shining white hairs; bracteoles similar, pilose chiefly along the midrib or margins. Flowers sessile. Tepals green, rather blunt, prominently 3-nerved with the nerves confluent exactly at the apex but not excurrent in a mucro, furnished with flexuose, multicellular, delicate smooth and shining whitish hairs; outer 2 tepals lanceolate-oblong, 3–3.5 mm., the nerves almost parallel and thus distant from the margins below and submarginal above, hyaline margins narrow; inner 2 broadly ovate with very wide hyaline margins on both sides (twice as wide as the nerved central portion at halfway up the tepal), pilose only along the dorsal surface of the midrib; middle tepal with one side broadly hyaline-margined and glabrous as with the inner 2, the other side resembling the outer 2. Stamens ± 1.75–2 mm., very delicate; pseudostaminodes none. Style long and slender, ± 1 mm. Capsule ovoid, 2–2.25 mm. Seeds subglobose or obscurely trigonous, ±1.5 mm., dark brown or black, shining, feebly reticulate. Fig. 21/7.

KENYA. Northern Frontier Province: 13 km. E. of Tarbaj, Dec. 1971, *Bally & Radcliffe-Smith* 14527!; Turkana District: base of Turkwell Gorge, Aug. 1962, *J. Wilson* 1275A!; Tana River District; 16 km. S. of Garissa near Tana R. bank, Jan. 1961, *Lucas* 56!

DISTR. **K**1, 2, 7; upper Egypt, tropical Arabia, Sudan, Ethiopia, Somalia

HAB. Open *Acacia* and *Acacia-Commiphora* bushland on red sandy soil; 140–790 m.

SYN. *Psilostachys gnaphalobryum* Hochst. in Flora 27, Beil.: 6 (1844)
Psilotrichum cordatum Moq. in DC., Prodr. 13 (2): 280 (1849); P.O.A.C: 173 (1895); F.T.A. 6(1): 60 (1909), *nom. superfl.* Type: as for *P. gnaphalobryum*
P. villosiflorum Lopr. in E.J. 27: 59 (1899) & in Malpighia 14: 452 (1901); F.T.A. 6(1): 61 (1909). Type: Ethiopia, between Mt. Robe and R. Daua, *Riva* 1455 (FI, holo.!)

2. **P. sericeum** *(Roxb.) Dalz.* in Dalz. & Gibson, Bombay Fl.: 216 (1861); Verdc. in K.B. 17: 491 (1964). Type: *Herb. Smith* 424.4 (LINN, lecto.!)

Erect or occasionally prostrate or ascending annual herb, ± 0.3–1 m. tall, much branched in the lower half and frequently also above, the lowest internodes of particularly the lower branches very divergent. Stem and branches striate, the lower internodes terete, the upper subquadrate and ± sulcate, nodes somewhat swollen; indumentum very variable, from subglabrous or with sparse soft pubescence to dense with a double indumentum of shorter soft hairs and long fine multicellular spreading bristly hairs (even when absent elsewhere tufts of bristly hairs occur between the branches and leaves at least of the upper nodes, frequently forming a ring there). Main stem and branch leaves cordate-ovate to broadly ovate or rarely cordate-lanceolate, obtuse to acuminate at the apex, 1.3–7 × 0.7–5.2 cm., the lowest on petioles up to ± 1(–1.8) cm. long, leaves towards the ends of the stem and branches reducing in size and sessile; indumentum very variable, from green on both surfaces (moderately softly appressed pilose below and thinly so with long and shorter multicellular strigose hairs above) to moderately appressed pilose above and densely silvery sericeous on the lower surface. Inflorescence a large open compound panicle of axillary opposite panicles formed of simple or branched spikes, each with a flexuose glabrous rhachis; peduncle and branches of the partial inflorescences capillary, glabrous or with long strigose multicellular hairs; bracts lanceolate, 0.75–1 mm., sparingly pilose, mucronate with the excurrent midrib; bracteoles similar but slightly shorter. Flower sessile. Tepals green, acute, prominently 3-nerved with the nerves confluent at the apex and excurrent in a minute mucro, glabrous or usually ± furnished with spreading white multicellular minutely denticulate hairs; 2 outer 2–2.5 mm., narrowly hyaline-margined with all 3 nerves equally strong, if hairs present then all nerves white pilose, hairs sometimes also present on the surface between; inner 2 widely hyaline-margined (the margin wider than the nerved central portion at halfway up the tepal), pilose only along the midrib, which is much stronger than the lateral nerves; middle tepal with one side as in the outer 2 and one as in the inner 2. Stamens delicate, ± 1.5–1.75 mm., pseudostaminodes none. Style short, ± 0.5 mm. Capsule ovoid, ± 1.75–2 mm. Seed subglobose, black, shining, feebly reticulate, ± 1.5 mm. Fig. 21/1-6.

KENYA. Northern Frontier Province: Korokoro [Korokora], N. bank of Garissa, June 1960, *Paulo* 468!; Mombasa, Mar. 1876, *Hildebrandt* 1985!; Lamu District: Kui I. Sept., 1956, *Rawlins* 120!

TANZANIA. Tanga, July 1953, *Drummond & Hemsley* 3230!; Pangani District: Mwera, Ushango, Sept. 1957, *Tanner* 3697B!; Uzaramo District: Msasani, Nov. 1968, *Batty* 317!; Zanzibar I., Mazizini [Massazine], Apr. 1961, *Faulkner* 2799!; Pemba I., Ngezi Forest, July 1901, *Lyne* 108!

DISTR. **K**1, 7; **T**3, 6, 8; **Z**; **P**, Mozambique, Somalia, Socotra, India

HAB. Mostly on sandy soil near the shore, along roadsides, beneath palms, in abandoned cultivation, grassy places, etc. at low altitudes (0–15 m.) but also inland in grassland, *Acacia* bushland, etc. up to 320 m.

SYN. *Achyranthes sericea* Roxb., Fl. Indica, ed. Carey 2: 502 (1824)
Psilostachys sericea (Roxb.) Hook. f. in G.P. 3: 32 (1880)
P. boivinianum Baill. in Bull. Soc. Linn. Paris 1: 622 (1889); E.P.A.: 70 (1953). Type: Zanzibar, *Boivin* (P, holo., photo.!)

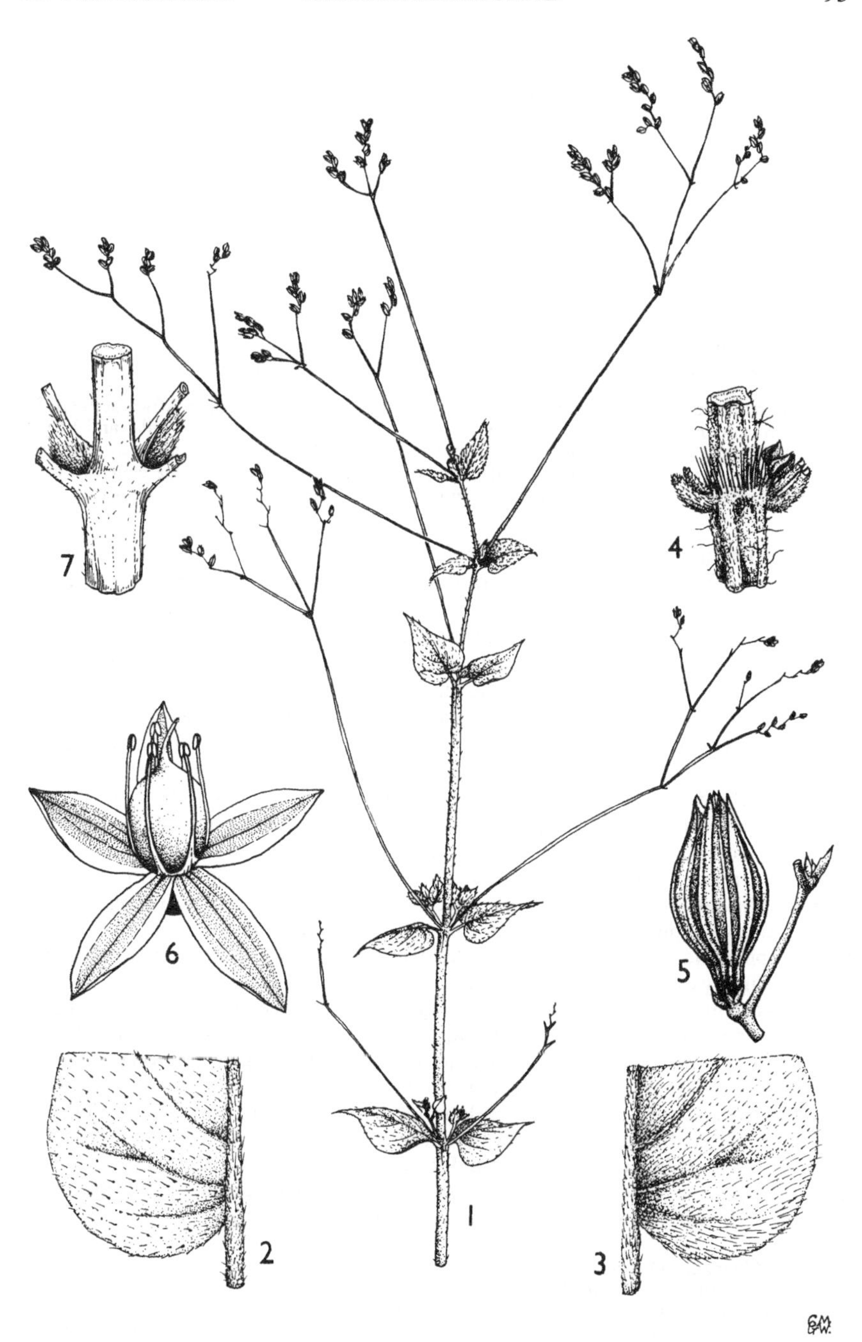

FIG. 21. *PSILOTRICHUM SERICEUM*—**1**, flowering branch, × $\frac{2}{3}$; **2**, upper surface of leaf, × 2; **3**, lower surface of leaf, × 2; **4**, node, × 4; **5**, flower, × 12; **6**, flower opened out, × 12. *P. GNAPHALOBRYUM*—**7**, node, × 4. 1–6, from *Polhill & Paulo* 546; 7, from *Gillett* 13096. Drawn by Christine Grey-Wilson.

P. nervulosa Baill., in Bull. Soc. Linn. Paris 1: 622 (1889); F.T.A. 6(1): 61(1909). Type: E. African coast, *Boivin* (P, holo., photo.!, K, iso.!)
P. filipes Baill. in Bull. Soc. Linn. Paris 1: 622 (1889).; F.T.A. 6(1): 61 (1909); F.D.O.-A. 2: 235 (1932) Type: Zanzibar, *Boivin* (P, holo., photo.!)
P. kirkii Bak. in K.B. 1897: 279 (1897). Type: Kenya, Kilifi District, Malindi, *Kirk* (K, holo.!)
Psilotrichum kirkii (Bak.) C.B. Cl. in F.T.A. 6(1): 60 (1909)
P. axillare C.B. Cl. in F.T.A. 6(1): 60 (1909); F.D.O.-A. 2: 235 (1932). Type: Kenya, Mombasa, *Hildebrandt* 1985 (K, holo.!)
P. edule C.B. Cl. in F.T.A. 6(1): 61 (1909); F.D.O.-A. 2: 235 (1932). Type: Pemba, *Lyne* 108 (K, lecto.!)
P. sericeovillosum Chiov. in Agric. Colon. 20: 103 (1926); E.P.A.: 71 (1953). Type: Somalia, Chisimaio, *Gorini* 28 (FI, lecto.!)
P. sericeovillosum Chiov. var *glabratum* Chiov. in Fl. Somala 2: 379 (1932) Type: Somalia, Jonti, *Gorini* 198 (FI, holo.!)
P. boivinianum (Baill.) Cavaco in Bull. Soc. Bot. Fr. 99: 184 (1952)

NOTE. As remarked by Verdcourt in K.B. 17: 491 (1964), intermediates occur in Africa between *P. sericeum* and *P. boivinianum*. Such intermediates also occur, however, in India, and I have been quite unable to separate the two species. In floral morphology they are identical, and leaf shape and indumentum are such unreliable characters in the Amaranthaceae generally (e.g. *Aerva, Achyranthes, Pupalia*) that I have no qualms about merging these two species.

3. **P. scleranthum** *Thwaites*, Enum. Pl. Zeyl.: 248 (1861); Verdc. in K. B. 17: 492 (1964). Type: Sri Lanka, Anuradhapura, *Gardner* (PDA, holo.!, K, iso.!)

Woody perennial herb or small shrub, erect or rooting at the basal nodes, sometimes ± scandent, 0.6–2 m., much branched with the branches spreading at 45–90°. Stem and branches in the older parts terete, striate and glabrescent, when young quadrangular and sulcate with pale thick corners, ± densely pilose with yellowish subappressed hairs, slightly swollen at the nodes. Leaves ovate to elliptic or elliptic-oblong, 2–10 × 1–4.6 cm., acute to acuminate at the mucronate apex, shortly to longly cuneate at the base with a 2–5 mm. petiole, moderately but finely pubescent on both surfaces (generally more conspicuously so on the lower surface of the primary venation). Inflorescences rather short spikes, 7–8 mm. wide and finally elongating to 2–5 cm., terminal and generally 2 at each node in the axils of the opposite leaves of stem or branches, sessile or on peduncles up to ± 4 cm. long; bracts ovate-lanceolate, 2–2.5 mm., ± densely appressed pubescent, sharply mucronate with the excurrent midrib, finally spreading or deflexed; bracteoles whitish, broadly deltoid-ovate with the margins slightly overlapping at the base, ±2 mm. with a distinct 0.5 mm. mucro formed by the excurrent midrib, glabrous or slightly pilose along the midrib and/or margins. Flowers sessile. Tepals white or greenish, very firm, faintly ± 5-nerved with still more obscure finer nerves; the 2 outer lanceolate-oblong, acute, sharply mucronate, 3.5–4.5 mm., finally indurate at the base, very narrowly hyaline-bordered, shortly pilose over the entire dorsal surface; the 2 inner lanceolate-ovate, shorter; acute to acuminate, pilose mainly centrally, the hyaline border widened below intermediate tepal hairy all over one side, the other broadly hyaline-margined. Stamens ±2.5 mm.; pseudostaminodes absent. Style ±0.75 mm. Capsule oblong-ovoid, ±2.5–3 mm. Seed ovoid, ± 1.75 mm., brown, shining, faintly reticulate. Fig 22/6.

KENYA. Machakos District: Kibwezi, Jan. 1906, *Scheffler* 98!; Kilifi District: Kikuyuni to Lake Jilore, Dec. 1954, *Verdcourt* 1186!; Lamu District: Utwani Forest, Dec. 1956, *Rawlins* 266!
TANZANIA. Mwanza District: Mbarika, Apr. 1953, *Tanner* 1368!; Tanga, Feb. 1893, *Holst* 2109!; Morogoro District: Turiani Falls, Mar. 1953, *Drummond & Hemsley* 1797!;? Zanzibar, 1854, *Boivin*! (may be coastal E. Africa)
DISTR. **K** 4, 7; **T** 1–8;? **Z**; eastern Zaire, Angola, Malawi, Mozambique, Zimbabwe, Madagascar, also S. India and Sri Lanka
HAB. Usually on forest fringes, in secondary cover beneath trees or in scrub, sometimes in open grassland, coconut plantations, roadsides or banks by streams; 0–1520 m.

SYN. *P. africanum* Oliv. in Hook., Ic. Pl. 16, t. 1542 (1886); P.O.A. C: 173 (1895); F.T.A. 6(1): 58 (1909); F.D.O.-A. 2: 233 (1932); Hauman in F.C.B. 2: 37 (1951); E.P.A.: 70 (1953);

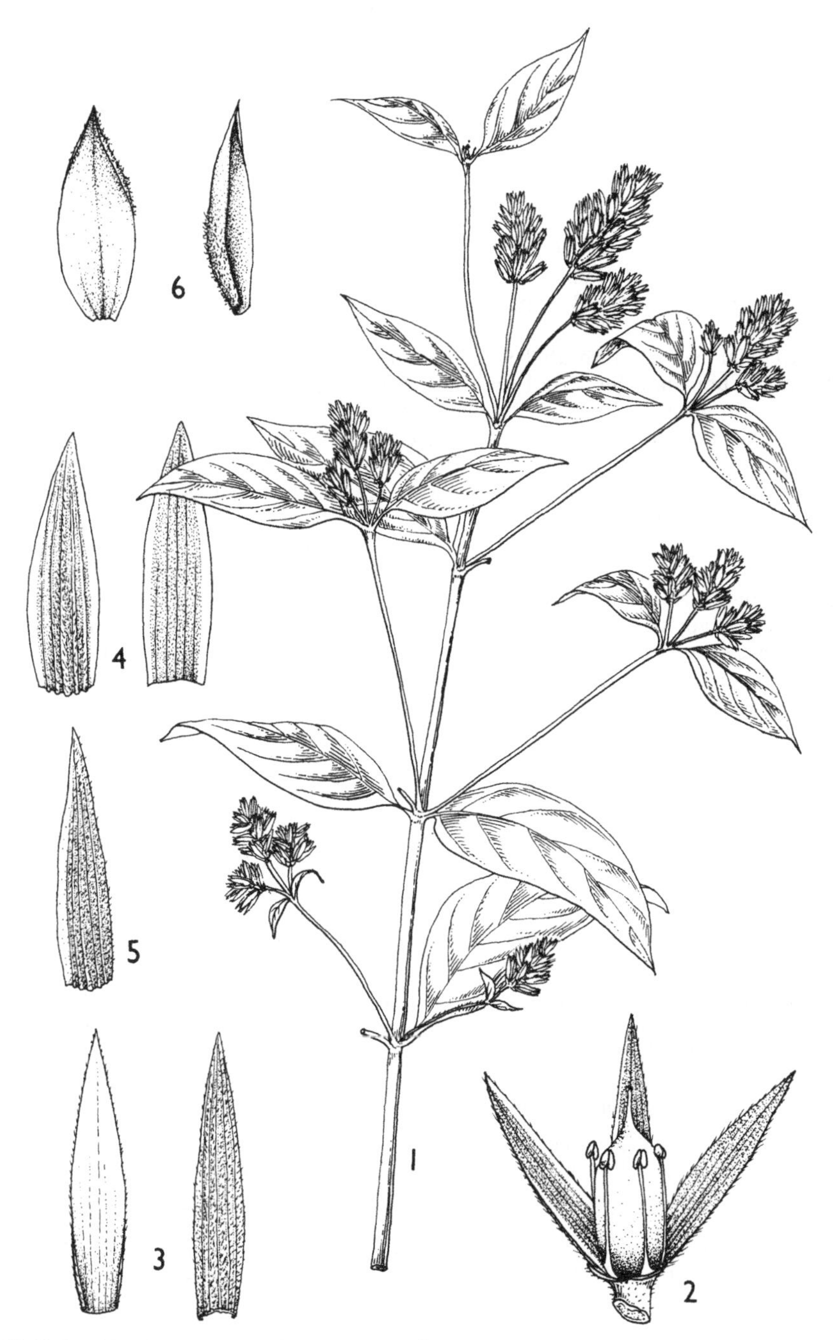

FIG. 22. *PSILOTRICHUM MAJUS*—**1**, flowering branch, × $\frac{2}{3}$; **2**, flower with two tepals removed, × 6; **3**, outer tepals, × 6; **4**, inner tepals, × 6; **5**, intermediate tepal, × 6. *P. SCLERANTHUM* —**6**, outer tepals, × 6. 1–5, from *Chandler* 1973; 6, from *Tanner* 1183. Drawn by Christine Grey-Wilson.

Cavaco in Mém. Mus. Nat. Hist. Nat. Paris, sér. B, 13: 111 (1962). Type: Tanzania, Kilimanjaro, *Johnston* (K, lecto.!)
P. africanum Oliv. var. *debile* Schinz in E.J. 21: 185 (1895); F.D.O.-A. 2: 234 (1932). Type: Malawi, Blantyre, *Last* (K, iso.!)
P. trichophyllum Bak. in K.B. 1897: 279 (1897); F.T.A. 6(1): 58 (1909); F.D.O.-A. 2: 234 (1932). Type: Mozambique, Shamo, *Kirk* (K, holo.!)
P. concinnum Bak. in K.B. 1897: 279 (1897); F.T.A. 6(1); 58 (1909). Type: Malawi, Blantyre, *Last* (K, holo.!)

4. **P. majus** *Peter* in F.R. Beih. 40(2), Descr.: 27, t. 27/1 (1938); Hauman in F.C.B. 2: 37 (1951). Type: Tanzania, Lushoto District, Sigi R. & Kwamtili, *Peter* 19933 & 25187 (B, syn.†)

Perennial herb or subshrub, 0.9–1.5 m. tall, branched from the base upwards with divaricate branches. Stem and branches in the older parts terete, thinly hairy or glabrescent, when young quadrangular, sulcate and ± densely crisped pilose, slightly swollen at the nodes. Leaves lanceolate to elliptic-oblong, 4–14 × 1.5–5 cm., acute or usually distinctly acuminate at the mucronate apex, at the base cuneate into the 3–7 mm. petiole, thinly and finely pubescent on both surfaces (but especially the lower surface of the primary venation) or subglabrous. Inflorescences of spikes or panicles, axillary or terminal, up to ± 10 cm. long, on peduncles 1–2.5 cm. long; bracts lanceolate, 2.5–3 mm., moderately pilose, sharply mucronate with the excurrent midrib, finally spreading; bracteoles whitish, broadly deltoid-ovate, with the margins slightly overlapping at the base, 2–3 mm., glabrous or finely ciliate, sharply mucronate with the excurrent green midrib. Flowers sessile or shortly pedicellate. Tepals green or greenish white, very firm, very strongly and prominently nerved, nerves confluent near the sharply mucronate apex; the outer 2 lanceolate-oblong, acute, 5.5–7 mm., finally indurate at the base, without or with a very narrow hyaline border, closely 7–9-nerved with dense fine appressed hairs chiefly somewhat divergent from the sides of the nerves; the inner 3 similar but slightly shorter, 5–7-nerved, more widely hyaline-margined (the outer of the three only along one side) and the margins glabrous. Stamens 2.5–2.75 mm., firm, without pseudostaminodes. Style 0.5–0.75 mm. Capsule ovoid, 3.5–4.5 mm. Seed ovoid, 2.5–3.5 mm., brown, shining, faintly reticulate. Fig. 22/1–5.

UGANDA. Bunyoro District: Budongo Forest, Sept. 1933, *Eggeling* 1426!; Busoga District: forest between Kamigo and Nakalange Hills, 22 km. NE. of Jinja, Oct. 1952 *G.H. Wood* 455!; Mengo District: Kajansi Forest, Oct. 1937, *Chandler* 1973!
KENYA. Kwale District: Shimba Hills, Lango ya Mwagandi [Longo mwagandi] area, Feb. 1968, *Magogo & Glover* 79! & Mar. 1968. *Magogo & Glover* 261! & Kwale Forest area, Mar. 1968, *Magogo & Glover* 428!
TANZANIA. Tanga District: Mt. Mlinga, Feb. 1917, *Peter* 19480!
DISTR. U2–4; K7; T3; eastern Zaire
HAB. Forest floor and margins; 440–1450 m.

SYN. *P. africanum* Oliv. forma *intermedium* Suesseng. in Mitt. Bot. Staats., München 1: 7 (1950). Type: Uganda, Mengo District, Mabira Forest, *Harris* 6 (K, holo.!)

NOTE. Both syntypes of this species were destroyed during World War II. The above-cited extant specimen (*Peter* 19480) was however cited in Peter's enumeration [Fedde Rep., Beih. 40(2): 235 (1938)], and was the specimen used for the preparation of his figure of *P. majus*. One syntype was cited as 19933 in the enumeration and 10933 in the description; the former is believed to be correct.

5. **P. fallax** *C.C. Townsend* in Publ. Cairo Univ. Herb. 7–8: 66 (1977). Type: Tanzania, Tanga District, Kange Gorge, *Faulkner* 1871 (K, holo.!)

Decumbent gregarious perennial herb, ± 0.3 m., sparingly branched from about the middle. Stem and branches terete, slender, densely furnished with crisped whitish hairs, somewhat thickened at the nodes. Leaves ovate to elliptic, 3–7.5 × 1.8–3.5 cm., blunt to subacute at the mucronate apex, shortly cuneate at the base, with a 2–4 mm. petiole, subglabrous on both surfaces. Inflorescences axillary, borne from near the base of the stem, of rather short spikes ± 1 cm. wide and finally elongating to ± 4 cm. long; axis densely whitish pilose; peduncles 3–8 mm.; bracts lanceolate, 2–3 mm., glabrous or almost so, sharply mucronate with the excurrent midrib, finally spreading; bracteoles broader, deltoid-ovate, 2–3 mm., greenish white to brownish, mucronate with the excurrent midrib, glabrous or sparingly ciliate. Flowers sessile. Tepals greenish white to brownish, lanceolate; outer 2±5 mm., strongly 5-nerved, with the surface sulcate between the nerves, moderately furnished with patent fine white hairs; inner 2 slightly shorter, 3(–4)-nerved, very sparsely pilose or glabrous, broadly hyaline-margined; intermediate tepal similar with only one hyaline margin; midrib and lateral nerves confluent at the apex in an indistinct mucro. Stamens short and delicate, ±1.75 mm.; pseudostaminodes absent. Ovary shortly pyriform, ±1 mm.; style short, 0.5–1.5 mm. Capsule ovoid, 3 mm. Seed ovoid, 2.75 mm., brown, shining, faintly reticulate.

KENYA. Kwale District: Shimba Hills, Mar. 1968, *Magogo & Glover* 414!; Kilifi District: Cha Simba, Jan. 1973, *Adams* 71!
TANZANIA. Tanga District: Kange Gorge, 8 June 1956, *Faulkner* 1871! & Pande, 19 Aug. 1982, *Hawthorne* 1464!
DISTR. **K** 7; **T**3; not known elsewhere
HAB. Shady forests on limestone; 0–380 m.

6. **P. axilliflorum** *Suesseng.* in B.J.B.B. 15: 69 (1938); Hauman in F.C.B. 2: 37 (1951) Type: Zaire, Avakubi, *Bequaert* 1826 (BR, holo.!)

Perennial herb, erect or decumbent and rooting at the lower nodes, simple or sparingly branched, 15–50 cm. Stem and branches terete, slender, densely furnished with crisped white hairs, often somewhat thickened at the nodes. Leaves elliptic, 3.5–9 (–12) × 1.25–4 cm., acute to slightly acuminate at the mucronate apex, long-cuneate at the base, slightly pilose on both surfaces to moderately appressed-pilose below. Inflorescences axillary, borne from near the base of the stem, of short few-flowered spikes ± 1–1.5 cm. wide and finally elongating to ± 2–7 cm. long, axis ± densely white pilose; peduncles 3–9 mm.; bracts lanceolate to lanceolate-ovate, ± 1.75 mm., scarious with a greenish centre, ciliate, sharply mucronate with the aristate midrib; bracteoles deltoid-ovate, ± 2 mm., green centrally with scarious margins, ciliate, mucronate with the excurrent midrib. Flowers sessile. Tepals greenish white, lanceolate, sharply acuminate; outer 2 tepals ± 7 mm., 5-nerved, the dorsal nerves very prominent and carinate, extending the full length of the tepal, the lamina between the ribs pale and pellucid, the 2 submarginal nerves less prominent, uniformly short-pilose, hyaline margins strongly involute; inner 2 tepals similar but slightly shorter, with no submarginal ribs and the hyaline margins densely furnished with long, ascending, multicellular white hairs; intermediate tepal with one side with a submarginal nerve and short hairs, the other side with no submarginal nerve and long-pilose; midrib and lateral nerves of all tepals confluent at the apex in an indistinct mucro. Stamens short and slender, ± 2–3 mm.; pseudostaminodes absent, or occasionally a minute tooth present in the sinus between the filaments. Ovary pyriform, ± 1.25 mm.; style rather slender, 1.25–1.5 mm. Ripe capsule and seed not seen.

UGANDA. Bunyoro District: NE. Budongo Forest, Aug. 1972, *Synott* 1112!
DISTR. U2; Zaire
HAB Ground-herb under old *Cynometra* forest in deep shade; ± 1000 m.

7. **P. elliotii** Bak. in F.T.A. 6(1): 58 (1909), as "*elliottii*" sphalm.; F.D.O.-A.

2: 234 (1932); U. K. W.F.: 137 (1974). Type: Uganda, E. side of Lake Edward, *Scott Elliot* 8062 (K, holo.!)

Perennial herb, somewhat woody at the base, procumbent and forming mats up to ± 1.2 m. across or scrambling among low vegetation, considerably branched from the base and often also above. Stem and branches in the older parts terete, striate, finally glabrescent at the base, quadrangular and sulcate when young, densely furnished with long white hairs, not swollen at the nodes. Leaves ovate to broadly oblong, 1–5 × 0.8–2.75 cm., rounded to shortly acuminate at the mucronate apex, rounded-subtruncate to shortly cuneate at the base, subsessile and semi-amplexicaul or with a petiole up to ± 5 mm. long, moderately or rarely densely furnished on both surfaces with rather long subappressed white hairs, sometimes glabrescent above with age. Inflorescences rather short spikes 6–8 mm. wide and finally elongating to 1.5–3 cm., terminal and generally 2 at each node in the leaf-axils, sessile or on very slender peduncles up to 3 cm. long; bracts deltoid-ovate, 2–2.5 mm., densely white pilose, finally spreading or deflexed, whitish with the only minutely excurrent midrib bordered by a conspicuous green vitta; bracteoles broadly cordate-ovate, 1 mm. long and 2 mm. wide, colour and indumentum similar to the bracts, similarly shortly mucronate. Flowers sessile. Tepals membranous, strongly 3-ribbed and greenish centrally with broad pale margins; the 2 outer lanceolate to oblong-lanceolate, ± 3–4 mm., with broader somewhat involute pale margins almost or quite as wide as the ribbed central part, the 3 ribs meeting below the apex, which is mucronate with the excurrent midrib, the pale margins uniformly furnished with ± patent white hairs; the inner 2 similar or more cucullate and indistinctly mucronate above, more narrowly pale-margined, densely long-pilose along both margins; the intermediate tepal long-pilose along one margin, the second margin with an indumentum similar to that of the outer tepals. Stamens ± 1.75 mm., very delicate; pseudostaminodes shortly deltoid to obsolete. Style ± 0.75 mm. Capsule oblong-ovoid, ± 1.5 mm. Seed ovoid, 1 mm., compressed, brown, shining, faintly reticulate.

UGANDA.. Toro District: Ruwenzori, Kikorongo, Dec. 1925, *Maitland!*; Ankole District: Kiruhura Jan. 1956, *Harker* 172!; Masaka District: near Kirumba, May 1972, *Lye* 6762!
KENYA. Machakos District: Makipenzi, June 1954, *Bally* 9749!; Kisumu, Feb. 1915, *Dummer* 1826!; Masai District: Aitong, Apr. 1961, *Glover, Gwynne & Samuel* 674!
TANZANIA. Mwanza District: Igalukiro, July 1973, *Tanner* 1594!; Lushoto District: Segoma, Mar. 1970, *Faulkner* 4330!; Morogoro District: ± 26 km. E. of Morogoro, Nov. 1955, *Milne-Redhead & Taylor* 7424!
DISTR. U1–4; K2, 4–6; T1–3, 5, 6; Zaire, Ethiopia, S. India and Sri Lanka
HAB. Usually in dry grassland or thin thornbush or on grassy or stony road verges, also scrambling among scrub on rocky slopes, in dried-up river beds, in secondary *Branchystegia* woodland and thinly grassed forest floor, on river banks and anthills; 0– ± 1700 m.

SYN. *Ptilotus ovatus* Moq. in DC., Prodr. 13(2): 281 (1849). Type: India, Madras, Wight (K, lecto.!)
Psilotrichum calceolatum Hook., Fl. Brit. Ind. 4: 725 (1885), *nom. illegit.* Type: S. India, Mt. Thomas, *Wallich* 6927A (K, lecto.!)
P. mildbraedii Schinz in Viert. Nat. Ges. Zürich 57: 553 (1912); F.D.O.-A. 2: 235 (1932). Type: Rwanda, *Mildbraed* 681 (B, holo.)
P. ovatum Peter in F.R. Beih. 40(2), Descr.: 28 (1932). Type: Tanzania, Mbulu District, Bonga-Bereku [Bereu], *Peter* 44174 (B, holo., photo.!)
P. ovatum (Moq.) Hauman in B.J.B.B. 18: 11 (1946); F.P.N.A. 1: 137 (1948); F.C.B. 2: 38 (1951), *nom. illegit.*
P. africanum Oliv. var. *pilosum* Suesseng. in K.B. 4: 478 (1950). Type: Uganda, Masaka District, Kabula, *Purseglove* 1824 (EA, K, iso.!)
P. peterianum Suesseng. in Mitt. Bot. Staats., München 1: 131 (1952). Type: as for *P. ovatum* Peter
[*P. nudum* sensu Cufod., E.P.A.: 71 (1953), *non* (Wall.) Moq.]
P. moquinianum Abeywickr. in Ceylon Journ. Sci., Biol. Sci. 2: 158 (1959). Type: as for *P. calceolatum*

NOTE. The specific epithet *calceolatum*, used by many authors for this species, is rendered illegitimate by the fact that when Hooker published it he cited *Ptilotus ovatus* Moq. as a synonym and thus should have taken up the epithet *ovatus* for the species. Hauman eventually did this, but by that time the name *Psilotrichum ovatum* (Moq.) Hauman was illegitimate as a later homonym. See Verdcourt in K.B. 17: 492–493 (1964).

8. **P. cyathuloides** *Suesseng. & Launert* in Mitt. Bot. Staats., München 2: 71 (1955). Type: Tanzania, Tanga District, Perani Forest, *Drummond & Hemsley* 3720 (K, holo.!, EA, iso.)

Perennial, 15–50 cm., much branched from the base upwards, with numerous arching stems from a central woody rootstock. Stem and branches striate, scarcely swollen at the nodes, in the older parts terete and glabrescent, increasingly sulcate and finally ± tetragonous above, the younger parts ± densely shortly pubescent. Leaves elliptic or ovate, 9–17 × 6–10 mm., obtuse at the apex and mucronate with the excurrent nerve, broadly cuneate at the base with a petiole 1–2 mm. long, ± densely appressed pubescent on both surfaces, only the midrib distinct. Inflorescences of short subsessile axillary and terminal spikes finally elongating to ± 1.5 cm., rhachis densely shortly pubescent; bracts lanceolate-ovate, 1–1.5 mm., shortly pilose, membranous except for the green midrib, which is excurrent in a short mucro; bracteoles similar but slightly broader and less pilose. Flowers sessile. Tepals green, rather firm, prominently 5-nerved, all the nerves confluent well below the apex to form a thick acumen which is mucronate at the tip; the outer 2 lanceolate, very narrowly hyaline-margined, 2.5–3.5 mm., densely shortly pubescent; the inner 2 ovate-lanceolate, widely hyaline-margined, pilose only at the apex and on the dorsal surface of the midrib; middle tepal with one side glabrous and widely margined, the other narrowly margined and uniformly pilose. Stamens delicate, ± 1.5 mm.; pseudostaminodes absent. Style ± 0.5 mm. Capsule oblong-ovoid with a small firm apex, ± 1.75 mm.; at maturity the fruiting perianth develops a hollow stipe to ± 2 mm. long, which is sulcate when it shrinks on drying. Seed oblong-ovoid, ± 1.5 mm., brown shining, faintly reticulate.

KENYA. Lamu District: Kui I., Mar. 1957, *Rawlins* 394!
TANZANIA. Tanga District: Perani Forest, Aug. 1953, *Drummond & Hemsley* 3720!; Uzaramo District: University College, Dar es Salaam, 1968, *Mwasumbi* 10335!; Kilwa District: Selous Game Reserve, Kingupira, Jan. 1978, *Vollesen* in *MRC.* 4836!
DISTR. **K**7; **T**3, 6, 8; not known elsewhere
HAB. Open grassy area in forests, exposed site on a low-lying island near the eastern coast, disturbed places; 0–150 m.

9. **P. vollesenii** *C.C. Townsend* in K.B. 34: 210 (1979). Type: Tanzania, Selous Game Reserve, *Vollesen* in *MRC.* 4485 (C, holo.!)

Perennial, ± 10 cm. tall, erect, branched from the base upwards. Stem and branches terete, striate, densely furnished with upwardly appressed whitish hairs. Leaves elliptic to ovate, 1.3–3.3 × 0.8–1.3 mm., subacute at the apex, shortly or more longly cuneate at the base with a very short petiole, moderately pilose on both surfaces with subappressed hairs. Inflorescence a short few-flowered spike, ± 1–2 × 1 cm.; bracts narrowly lanceolate, 2–2.5 mm., green with a narrow hyaline border below, shortly pilose; bracteoles much broader, deltoid-ovate, 2–2.25 mm., hyaline, the greenish midrib excurrent in a sharp arista, glabrous. Flowers sessile. Tepals green, very narrowly acuminate, firm, lanceolate with 5 thick whitish nerves which are confluent above and run out into a very fine short arista; the outer 2 extremely narrowly hyaline-margined, 6 mm., densely furnished with rather stiff shortish hairs over the entire surface; inner 2 similar but 1 mm. shorter and hairy only along the midrib; middle tepal intermediate in length, pilose on one side only. Stamens delicate, ± 2.5 mm. long; pseudostaminodes absent. Style slender, 1.5 mm. Capsule oblong-ovoid with a small firm apex, 2.75 mm. long. Seeds oblong-ovoid, 2.5 mm., brown, somewhat shining, feebly reticulate.

TANZANIA. Kilwa District: Selous Game Reserve, Nunga, Feb. 1977, *Vollesen* in *MRS.* 4485!
DISTR. **T**8; not known elsewhere
HAB. *Manilkara sulcata, Hymenaea verrucosa, Neogordonia parvifolia,* etc. thicket; ± 700 m.

NOTE. At present this can only be regarded as distinct; it is only known from the holotype, and little material of the undoubtedly very close *P. cyathuloides* has been seen. The differentiating characters between these two needs further testing.

10. **P. schimperi** *Engl.*, Hochgebirgsfl. Trop. Afr.: 207 (1892); F.T.A. 6(1): 59 (1909); F.P.S. 1: 120 (1950); E.P.A.: 71 (1953); U.K.W.F.: 136 (1974). Type: Ethiopia, R. Reb, *Schimper* 1863, No. 1388 (EA, K!, iso.)

Erect or decumbent annual, (15–) 30–75 cm. tall, with numerous branches from the base upwards (or small plants unbranched below the inflorescence), vegetative parts glabrous or with scattered strigose hairs (especially along the margins and midrib of the upper leaves and on the upper internodes). Stem and branches strongly striate, subsulcate, or the basal internodes terete, nodes slightly swollen. Leaves linear, 3.5–12 × 0.2–1 cm., rather blunt at the apex with a distinct mucro formed by the confluence of the thickened margins and strong excurrent midrib, attenuate and subsessile at the base. Inflorescences of axillary spikes ± 0.7 cm. wide and finally elongating to as much as ± 14 cm., on peduncles 0.2–6.5 cm. long, the upper usually alternate; bracts deltoid-lanceolate, 1.5–2 mm., membranous except for the greenish excurrent midrib, finally spreading; bracteoles similar but slightly smaller. Flowers sessile. Tepals green or sometimes suffused with red or brown, prominently 3-nerved and the nerves confluent with the midrib above but not excurrent in a mucro; outer 2 lanceolate-oblong, 3.5–4 mm., obtuse, with a very narrow hyaline border, shortly scabrid-pilose along the margins, nerves and often also between the nerves; inner 2 more broadly hyaline-bordered, the margins fringed with long white hairs; middle tepal with one margin broadly hyaline and long-pilose, the other narrowly hyaline and scabrid. Stamens ± 2.5 mm., very delicate, almost free; pseudostaminodes none. Style ± 0.5 mm. Capsule ovoid, 2.75–3 mm. Seed ovoid, 1.5–2 mm., black, shining, faintly reticulate.

UGANDA. Karamoja District: near Kotido, Sept. 1958, *J. Wilson* 617! & Kangole, Aug. 1972, *J. Wilson* 2096!
KENYA. Fort Hall District: Thika, N. of Thika R., Aug. 1967, *Faden* 67/634A!; Kiambu District: Thika town, July 1971, *Lye* 6345!; Nairobi District: 19 km. NE. of Nairobi, Jan. 1952, *Bogdan* 3389!
TANZANIA. Shinyanga District: near Shinyanga, *Bax* 45!; Arusha, lower Nduruma R., June 1928, *Haarer* 1417!; Dodoma District: 33 km. W. of Itigi Station, Apr. 1964, *Greenway & Polhill* 11571!
DISTR. **U**1; **K**3, 4; **T**1, 2, 4, 5, 8; Ethiopia, Zambia, Rwanda
HAB. Usually on damp ground; in ditches, seasonally wet grassland, on stream banks, in *Acacia* woodland – even in deep mud in a swamp; soil black clay: 110–1800 m.

SYN. *P. angustifolium* Gilg in N.B.G.B. 1:328 (1897); F.T.A. 6(1): 61 (1909); F.D.O.-A. 2: 235 (1932). Type: Tanzania, *Stuhlmann* 3470 (B, holo.!)
P. camporum Hauman in B.J.B.B. 18: 110 (1946) & in F.C.B. 2: 39 (1951). Type: Rwanda, between Murwita and Kansekaba, *Lebrun* 9707 (BR, holo.!)
P. gramineum Suesseng. in Mitt. Bot. Staats. München 1: 111 (1952). Type: Kenya, 3 km. N. of Nairobi, *Bogdan* 1642 (K, holo.!)
P. schimperi Engl. var. *gramineum* (Suesseng.) Suesseng. in Mitt. Bot. Staats., München 1: 194 (1953)

18. ACHYRANTHES

L., Sp. Pl.: 204 (1753) & Gen. Pl., ed. 5: 96 (1854)

Herbs with opposite, petiolate, entire leaves. Inflorescence a ± slender spike, terminal on the stem and branches, the flowers at first congested and ± patent, finally usually laxer and deflexed; bracts deltoid or ovate, the midrib excurrent in a spine. Flowers solitary in the bracts, ⚥, bibracteolate. Perianth-segments 4–5, 1–3(–5)-nerved, narrowly lanceolate, acuminate, aristate with the excurrent midrib, indurate in fruit especially at the base. Bracteoles spinous-aristate with the excurrent midrib, the lamina forming short and free to longer and adnate membranous wings. Stamens 2–5; filaments filiform, monadelphous, alternating with quadrate to broadly quadrate-spathulate pseudostaminodes, these simple and dentate or fimbriate, or furnished with a variably developed dorsal scale; anthers bilocular. Ovary with a solitary pendulous ovule, the ovary wall very thin in fruit; style slender; stigma small, truncate-capitate. Entire flower with bracteoles falling with the ripening of the cylindrical seed, the deflexed bracts persistent. Endosperm copious.

6 species in the warm temperate and tropical regions of the world.

Very close to *Pandiaka*, but with a very characteristic appearance owing to the generally elongate and rather lax inflorescence, the narrow glabrous perianth with obscure nerves, the bracteoles with a relatively short lamina and a dorsally prominent midrib excurrent in a long, terete arista, and the perianth and bract commonly deflexed and closely appressed to the inflorescence axis in ripe fruit.

Inflorescence not fasciculate, the spikes solitary, axillary and terminal, regularly disposed .... 1. *A. aspera*
Inflorescence fasciculate, the spikes stellately clustered at the ends of the main stem and branches .... 2. *A. fasciculata*

1. **A. aspera** *L.*, Sp. Pl.: 204 (1753); P.O.A.C : 173 (1895); F.T.A. 6(1): 63 (1909); F.D.O.-A. 2: 237(1932); F.P.N.A. 1: 135 (1948); F.P.S. 1: 113 (1950); Hauman in F.C.B. 2: 53 (1951); E.P.A.:72 (1953); Cavaco in Mém. Mus. Nat. Hist. Nat. Paris. sér. B, 13: 114 (1962); F.P.U.: 101 (1962); U.K.W.F.: 137 (1974). Type: Sri Lanka, *Hermann Herbarium* Vol. 2, p. 69, right-hand specimen (BM, lecto.!)

Perennial herb (sometimes woody and somewhat suffrutescent), occasionally flowering in the first year, 0.2–2 m., stiffly erect to subscandent or straggling and ± prostrate, simple to much branched, stems stout to very weak, distinctly to obscurely 4-angled, striate or sulcate, subglabrous to densely tomentose, the nodes ± shrunken when dry. Leaves elliptic, oblong or ovate, acute or acuminate to almost round and very obtuse, gradually or abruptly narrowed below,(2–)5–22(–28) × 1.3–8(–10) cm., indumentum varying from subglabrous through subglabrous above and densely appressed-canescent below to ± densely tomentose on both surfaces; petioles of main stem leaves 3–25(–30) mm., shortening above and below. Inflorescences at first dense, finally elongating to (5–)8–34(–40) cm., peduncles (0.6–)1–6(–7.5) cm. Bracts lanceolate or narrowly deltoid-lanceolate, pale or brownish-membranous, 1.75–5(–6) mm., glabrous. Bracteoles 1.5–4.5(–6) mm., the basal wings ⅓ – ¼ the length of the spine and ± adnate to it (sometimes free above or tearing free), typically tapering off above but not rarely rounded or truncate. Perianth whitish or pale green to red or purple; segments 5, 3–7(–10) mm., the outer pair longest, narrowly lanceolate to lanceolate, very acute, with a distinct midrib and 2 obscure to distinct lateral nerves, narrowly or moderately pale-margined. Stamens 5, the filaments 1.5–4.5(–6) mm., alternating with subquadrate pseudostaminodes; typically the apex of the latter curves slightly inwards as a narrow, crenate or entire, often very delicate flap, while from the dorsal surface arises a fimbriate-ciliate scale extending across the width of the pseudostaminode; not rarely, however, this is reduced to a "stag's-horn" process at the centre of the dorsal surface or a shallow, dentate rim, or even becomes small and filiform–or else subapical or apical so that the pseudostaminode appears simple (this usually in small forms of var. *sicula*). Style slender, 1–4(–6) mm. Capsule oblong-ovoid, 1–3(–5) mm. Seed filling the capsule, oblong-ovoid, smooth.

KEY TO INFRASPECIFIC VARIANTS

Leaves broadly elliptic, broadly obovate to almost round,

blunt or abruptly apiculate; flowers (Fig. 23/3) short and rather plump, to ±4.5 mm. .... a. var. **aspera**
Leaves oblong to elliptic, bluntly to long-acuminate; flowers (Fig. 23/5) 3–7 mm., but less plump and if not exceeding 4.5 mm. the leaves frequently distinctly acuminate:
Leaves with variable indumentum but not greenish above and silvery canescent beneath; flowers mostly 5–7 mm b. var **pubescens**
Leaves distinctly acuminate, typically greenish above and, at least when young, silvery canescent beneath; flowers mostly 3–4.5 mm .... c. var. **sicula**

a var. **aspera**

Plant stout and robust, the leaves broadly elliptic or obovate to almost round, blunt or abruptly apiculate, frequently tomentose at least on the lower surface, especially along the veins. Flowers 3–4.5 mm., usually shorter and plumper than in var. *pubescens*. Fig. 23/1–3.

UGANDA. W. Nile District: between Omogo rest camp and R. Oru, Aug. 1953, *Chancellor* 155! & 20 km. W. of Pakwach, Oct. 1974, *Katende* 2338!
KENYA. Kisumu, Feb. 1915, *Dummer* 1842!; Tana River District: by Tana R. at Garsen, Mar. 1977, *Hooper & Townsend* 1163!
TANZANIA. Kwimba District: near Magu, Apr. 1953, *Tanner* 1517!; Lushoto District: Korogwe, Nov. 1962, *Archbold* 14!; Morogoro, June 1948, *Semsei* 2378!; Zanzibar I., Chukwani, Apr. 1961, *Faulkner* 2805!; Pemba I., Chake Chake, Sept. 1929, *Vaughan* 655!
DISTR. **U**1; **K**5, 7; **T**1, 3, 6, 8; **Z**; **P**; see below.

SYN. *A. aspera* L. var. *indica* L., Sp. Pl.: 204 (1753); E.P.A.: 73 (1953). Type: as for species
A. indica (L.) Mill., Gard. Dict., ed. 8 (1768)
A. obtusifolia Lam., Encycl. 1: 545 (1785). Type: source unknown, *Sonnerat* (P-LA, holo.! IDC microfiche 546. 11!)
A. obovata Peter in F.R. Beih. 40(2), Anh.: 25 (1932). Type: Tanzania, Lushoto District, Mashewa, *Peter* 13869 (B, holo.!)
A. aspera L. var. *obtusifolia* (Lam.) Suesseng. in Mitt. Bot. Staats., München 1: 152 (1952)

NOTE. Only one or two intermediates between this variety and var *pubescens* have been seen, e.g. *Semsei* 3091, from Mombo Forest (**T**3, Lushoto District). The African material as cited above is a good match for the Linnean type of the variety, and its distribution favours original presence along the coastal districts of East Africa, or introduction there from India. The inland localities are almost all in districts round the eastern shore of Lake Victoria, which again suggests distribution by trading.
In var. *aspera* the ventral scale of the pseudostaminodes is often very delicate and sometimes interrupted; but the dorsal scale is in my experience almost always considerably fimbriate.

b. var. **pubescens** *(Moq.) C.C. Townsend* in K.B. 29: 473 (1974). Type: Mexico, *Berlandier* 79, 104, 105 (all three numbers on a single specimen) (G-DC, lecto., IDC microfiche 2196. 20!)

Plant robust, the perianth-segments generally (4–)5–7 mm. Leaves oblong to elliptic, not appressed-canescent on the lower surface, subglabrous to pilose or tomentose, bluntly to distinctly acuminate. Fig. 23/4–8.

UGANDA. Karamoja District: at base of Mt. Debasien, Amaler Jan. 1936, *Eggeling* 2527!; Mbale District: Butandiga, Sept. 1932, *A.S. Thomas* 665!; Mengo District: Wabusana-Luwero, July 1956, *Langdale-Brown* 2252!
KENYA. Northern Frontier Province: Moyale, Dec. 1952, *Gillett* 15079!; Naivasha District: Lake Naivasha area, Aug. 1966, *E. Polhill* 192!; Masai District: Narosura Rest Camp, Aug. 1961, *Glover, Gwynne, Samuel & Tucker* 2431!
TANZANIA. Mwanza District: Nyegezi, Apr. 1952, *Tanner* 676!; Lushoto District: Magunga Estate, July 1952, *Faulkner* 994!; Iringa District: Ruaha National Park, 20 km. from Msembe on the Great Ruaha R., Aug. 1969, *Greenway & Kanuri* 13781!; Zanzibar I., Marahubi, July 1963, *Faulkner* 3209!
DISTR. **U** 1–4; **K** 1, 3–7; **T** 1–5, 7; **Z**; see below

SYN. *A. fruticosa* Lam. var. *pubescens* Moq. in DC., Prodr. 13(2): 314 (1849)
[*A. bidentata* sensu F.T.A. 6(1): 64 (1909); F.D.O.-A. 2: 236 (1932); Cavaco in Mém. Mus. Nat. Hist. Nat. Paris, sér. B, 13: 125 (1962); Townsend in K.B. 28: 145–6 (1973), quoad pl. afr. or., *non* Blume]
A. aspera L. var. *nigro-olivacea* Suesseng in F.R. 51: 194 (1942). Type: Tanzania, Ulanga District, Sali, *Schlieben* 2264 (B, holo.)

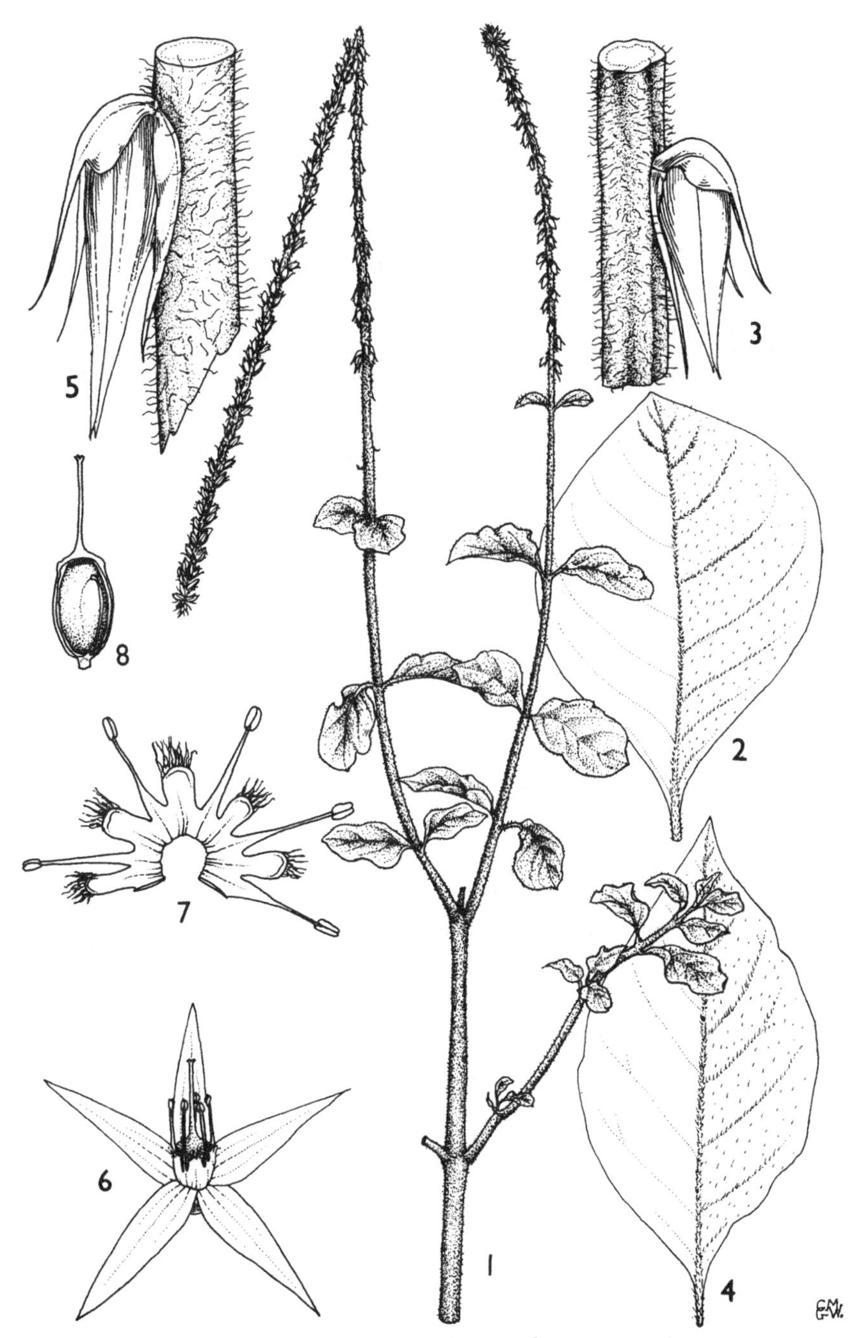

FIG. 23. *ACHYRANTHES ASPERA* var. *ASPERA*—**1**, flowering branch × $\frac{2}{3}$; **2**, single leaf, × $\frac{2}{3}$; **3**, flower on inflorescence axis, × 6. *A. ASPERA* var. *PUBESCENS*—**4**, single leaf, × $\frac{2}{3}$; **5**, flower on Inflorescence axis, × 6; **6**, flower opened up, × 4; **7**, androecium, × 6; **8**, gynoecium, × 6. 1, 3, from *Archbold* 14; 2, from *Faulkner* 3521; 4, from *Mathenge* 100; 5–8, from *Hancock* 121. Drawn by Christine Grey-Wilson.

A. aspera L. forma *robustiformis* Suesseng. in Mitt. Bot. Staats.,München 1: 70 (1950). Type: Tanzania, Masai Distirct, Ngorongoro, *Bally* 2273 (K, lecto.!)

NOTE. Suessenguth names some Ugandan (U4) specimens as var. *porphyrostachya* (Moq.) Hook.f., an almost glabrous form of this species described from India. None of these specimens is quite as slender in the inflorescence or as glabrous in the young leaves as the Indian plant-which I believe is in any case only a shade form of var. *pubescens*, of which it has quite the leaf-shape. The same observation applies to material so named by Cavaco (l.c.: 122, 1962) from Kenya and Tanzania. In Kew Bulletin 28: 145–6 (1973) I followed Suessenguth and Cavaco in referring some African plants to *Achyranthes bidentata* Blume. Further studies (particularly the opportunity of studying such forms in the field in the Kakamega region) have convinced me that Hauman, F.C.B. 2: 54(1951), was correct in including these forms within the spectrum of *A. aspera*, of which they are subglabrous forest forms with often very reduced (but constantly present) dorsal scales to the pseudostaminodes. The leaves are acute or shortly acuminate, not long-acuminate as in the Asiatic *A. bidentata*, and the flowers appear to deflex more rapidly. These African forest forms are very similar in general appearance to the var. *porphyrostachya* referred to above and, like it, grade into the normal hairier plants of disturbed ground, waysides, etc.

c. var. **sicula** *L.*, Sp. Pl.: 204 (1753); F.D.O.-A. 2: 238 (1932); Cavaco in Mém. Mus. Nat. Hist. Nat. Paris, sér. B, 13: 119 (1962). Type: "Amaranthus radice perpetua tab. 9", *Herb. Boccone* (P, lecto., photo.!)

Plant more slender, the perianth-segments generally 3–4.5(–5) mm. Leaves usually silvery canescent below, green above, rather long-acuminate.

UGANDA. Karamoja District: Moroto, Sept. 1956, *J. Wilson* 256!; Kigezi District: Echuya Reserve, near Kanaba Gap, Apr. 1950, *Dawkins* 576!; Elgon, Jan. 1918. *Dummer* 3486!
KENYA. Northern Frontier Province: Furrole Mt., Sept. 1952, *Gillett* 13954!; Uasin Gishu District: Kipkarren, *Brodhurst-Hill* 298!; Embu, Mar. 1932, *Sunman* 2213!
TANZANIA. Mwanza District: Busisi, Apr. 1952, *Tanner* 667!; Mpwapwa, May 1929, *Hornby* 127!; Rungwe Distirct: Mbeye, Mar. 1932, *St. Clair-Thompson* 808!
DISTR. **U** 1–4; **K** 1–7; **T** 1–8, see below

SYN. *A. argentea* Lam., Encycl. Méth. 1: 545 (1785). Type: cultivated material from Paris (P-LA, holo.!, IDC microfiche 546, 12!)
A. aspera L. var. *argentea* (Lam.) Boiss., Fl. Orient. 4: 994 (1879); F.T.A. 6(1): 63 (1909); F.D.O.-A. 2: 239 (1932); F.P.N.A. 1: 136 (1948); Hauman in F.C.B. 2: 55 (1951)
A. annua Dinter in F.R. 15: 82 (1917). Type: Namibia, Eahero, *Dinter* 3303 (B, holo.)
A. aspera L. forma *rubella* Suesseng. in B.J.B.B.15: 54 (1938). Type: Burundi, Kizoze, *Lejeune* 190 (BR, holo.)
A. aspera L. forma *annulosa* Suesseng. in Mitt. Bot. Staats., München 1: 69 (1950). Type: Uganda, Kigezi District, Ishasha Gorge, *Purseglove* 2284 (K, holo.!, EA, iso.) – shade form
A. aspera L. var. *sicula* L. forma *latifolia* Suesseng. in Mitt. Bot. Staats., München 1: 70 (1950). Type: Mt. Kenya, *Fries* 781 (K, holo.!)
A. argentea Lam. var. *annua* (Dinter) Suesseng. in Mitt. Bot. Staats., München 1: 152 (1952)
A. argentea Lam. var. *albissima* Suesseng. in Mitt. Bot. Staats., München 2: 70 (1955). Type: Kenya, Northern Frontier Province, Furroli, *Gillett* 13954 (K, holo.!, EA, iso.)

NOTE. Intermediates between this variety and var. *pubescens* are relatively frequent; the development of the dorsal scale of the pseudostaminode in this variety varies from normally developed as in var. *pubescens* to "stagshorn" or much more reduced, or sometimes subapical or apical so that the pseudostaminode appears simple.

Various insignificant formae and subformae of this variety were described by Suessenguth.

DISTR. (of species in general). Found practically throughout the world in tropical and warmer regions generally. Some forms may have originated in more restricted areas (e.g. var. *sicula* in the Mediterranean region, where it still appears to be the only form occurring, and var. *indica* in India), but these have now become sufficiently widely dispersed to make this a matter of conjecture
HAB. (of species in general). In practically all habitats from coastal scrub and dry cliff tops to mist forest; common as a weed of cultivation or disturbed ground, in open grassland, hardpan patches between rocks, along forest trails and edges, in seasonal swamps and dried-up watercourses, etc.; grows on a wide range of soil types from red sandy ground to black cotton soil or granite mountain tops; sea-level to 3000 m.

2. **A. fasciculata** *(Suesseng.) C.C. Townsend* in K.B. 34: 424 (1980). Type: Tanzania, Mbulu District, Nangwa, *Greenway* 7620 (K, holo.! EA, iso.)

Bushy or scandent herb, with numerous ± divaricate branches, 0.5–1.5 m. (probably more); stem and branches quadrangular with paler angles, moderately to densely brownish tomentose with ascending or upwardly appressed hairs. Leaves oblong-ovate to oblong-elliptic, 2–5 × 1.3–2.8 cm., dark green and moderately appressed pilose on the upper surface, whitish (especially when young) and ± densely appressed pilose beneath, acuminate, shortly cuneate to subtruncate at the base with a distinct petiole to ± 1.3 cm. long. Inflorescences terminal on the stem and axillary in the upper leaves by branch reduction, silvery and commonly ± pink-flushed, formed of fascicles of spikes of considerably dissimilar length, the longest to ± 4.5 × 1.5 cm., on a tomentose peduncle 1.5–6 cm. long, axis villous. Bracts membranous, lanceolate, 3–4.5 mm., finally deflexed, glabrous, long-aristate with the excurrent midrib. Bracteoles membranous lanceolate, 3.5–5 mm., glabrous or the margins shortly and finely ciliate, the arista formed by the excurrent nerve comprising half the length. Tepals glabrous, narrowly lanceolate; 2 outer straight or slightly curved at the tips, faintly 1–3-nerved, 5.5–8 mm., sharply pointed and shortly aristate with the ± perceptibly excurrent midrib, hyaline margins absent; 2 inner slightly shorter, 1-nerved, delicately and rather broadly hyaline-margined, narrowly tapering but not sharply pointed; middle tepal intermediate, 1-nerved, one margin hyaline. Perianth and bracteoles apparently falling together. Stamens 2.5–4 mm.; pseudostaminodes oblong-flabelliform. 1.5–2.5 mm., with an incurved dentate apical lobule and a densely long-fimbriate dorsal scale. Ovary obpyriform, ± 1.5 mm., with a firm, rounded apex; style slender, 3–3.5 mm. Mature fruit not seen.

TANZANIA. Masai District: Ngorongoro Crater, 15 Aug. 1952, *Tanner* 857! & Embagai, 5 Feb. 1932, *St. Clair-Thompson* 1414!; Mbulu District: Hanang, Dec. 1929, *B.D. Burtt* 2262!
DISTR. T2; not known elsewhere
HAB. Upland grassland and mist forest on volcanic soils; 1820–2730 m.

SYN. *Pandiaka fasciculata* Suesseng. in K.B. 4: 477 (1950); Cavaco in Mém. Mus. Nat. Hist. Nat. Paris, sér, B, 13: 148 (1962)

Doubtful species

A. pedicellata *Lopr.* in E.J. 27: 56 (1899). Type: Tanzania, Biharamulo District, Kimwani [Kimoani] Plateau, *Stuhlmann* 3390. No material of this was seen for F.T.A., and none exists at B, BM, FI, K, M or Z. From the original description it can hardly be an *Achyranthes* in the sense of the present account; but the description does not seem to fit any other East African amaranth known to me either.

A. viridis *Lopr.* in E.J. 27: 55 (1899). Type: Tanzania, Usambara Mts., Magamba Forest, *Holst* 3800. Not seen for F.T.A. and no material has been seen. The original description indicates a true *Achyranthes,* probably one of the greenish, rather thinly hairy forest forms of *A. aspera* var. *pubescens.*

A. winteri *Peter* in F.R., Beih. 40(2); 240 (1938). The very brief description of this species is quite non-diagnostic, and none of the three syntypes cited by Peter are available, being presumed destroyed at Berlin in World War II.

19. CENTROSTACHYS

Wall. in Roxb., Fl. Indica 2: 497 (1824)

Perennial herb with opposite leaves and branches. Leaves entire. Flowers bibracteolate in shortly pedunculate elongate bracteate spikes which are terminal on the stem and branches, each bract subtending a single flower; bracts persistent, hyaline; bracteoles round, hyaline, falling with the flower. Perianth-segments 5, somewhat spreading at anthesis, later closing together and considerably indurate at the base; upper tepal narrowest and longest, 1–3-nerved. Stamens 5, shorter than the perianth, shortly monadelphous at the base, alternating with spathulate pseudostaminodes furnished with fimbriate dorsal scales; anthers bilocular. Ovary with a single ovule pendulous on a curved funicle; radicle ascending; style filiform; stigma capitate. Capsule thin-walled, tightly enclosing the seed, falling together with the persistent perianth and bracteoles. Endosperm copious.

A monotypic genus.

C. aquatica *(R. Br.) Moq.* in DC., Prodr. 13(2): 321 (1849); F.P.S. 1: 118 (1950); Hauman in F.C.B. 2: 56 (1951); E.P.A.: 74 (1953). Type: Thailand, *Koenig* (BM, holo.!)

Aquatic or subaquatic perennial 0.5–1.5 m., prostrate to straggling or erect, usually much branched, considerably rooting at the lower nodes with dense tufts of whitish rhizoids; stem near the base up to 2 cm. thick, spongy, hollow. Upper stem and branches sulcate-striate, glabrous for the most part but increasingly appressed-pilose towards the inflorescence. Leaves lanceolate to lanceolate-oblong or oblong-ovate, cuneate or usually attenuate to the base, acute to acuminate at the apex, moderately appressed pilose on both surfaces (densely so when young); lamina (2.5–)7–15 × (0.8–)2–5 cm.; petiole 0.4–4 cm. Spike ± 4–12 cm. in flower, elongating to 25 cm. or occasionally even more in fruit; rhachis moderately to densely appressed pilose; peduncle short, mostly ± 1.5 cm. Bracts deltoid-lanceolate, 3–4 mm., hyaline (drying pale brownish) with a single midrib, glabrous, finally deflexed below the hard callus left by the fallen perianth; bracteoles round, hyaline, ± 1.5–2 mm., glabrous. Perianth 6–8 mm., the upper (outer) tepal, 1–3-nerved, slightly longer than the remainder, with a sharper often slightly recurved tip, and a somewhat narrower pale border; remaining tepals blunter, with up to 7 nerves. Filaments stout, 2–3 mm. Style 1.75–2.5 mm. Capsule ± 4 mm., slightly broader at the base but rounded above. Seed smooth, chestnut brown. Fig. 24.

UGANDA. Acholi District: Abera, Gulu, Nov. 1945. *A.S. Thomas* 4329!; Lango District: Cawente [Chiawante], July 1935, *H.B. Johnston* 1019!; Karamoja District: Lochomon R., July 1963, *J. Wilson* 1334!

KENYA. Tana River District: 7 km. E. of Garsen towards Witu, Mar. 1977, *Hooper & Townsend* 1227!

TANZANIA. Shinyanga District: Maliya, Aug. 1957, *Grundy* in *E.A.H.* 11397!; Masai District: Eluanata Dam, June 1946, *Greenway* 7791!; Dodoma District: Imagi Hill, Apr. 1962, *Polhill & Paulo* 2096!

DISTR. **U**1; **K**7; **T**1, 2, 4–7; Nigeria, Cameroun, Zaire, Burundi, Sudan, Ethiopia, Zambia, Mozambique, Zimbabwe, India, Sri Lanka, Burma, Thailand, Java

HAB. In pools and waterholes, along rivers and lake shores, and often about dams; on soils ranging from sand and grey-brown clay loam to black clay; 20–1460 m.

SYN. *Achyranthes aquatica* R.Br., Prodr. Nov. Holl.: 417 (1810); F.T.A. 6(1): 64 (1909); F.D.O.-A. 2: 239 (1932); Cavaco in Mém. Mus. Nat. Hist. Nat. Paris. sér. B. 13: 124 (1953)

NOTE. Although the authorities for the binomial *Centrostachys aquatica* have been generally accepted as (R. Br.) Wall. in Roxb., Fl. Ind. 2: 497 (1824), such citation is incorrect. The species is listed in that place under *Achyranthes* as "3. *A. (Centrostachys* Wall.*) aquatica,* R." Wallich then describes his new genus *Centrostachys* and states that he proposes detaching "this species" (i.e., *Achryanthes aquatica*) into it. Roxburgh does not appear to have accepted Wallich's genus, as the species reappears as an *Achyranthes* in Carey's edition of the Flora Indica 1: 673 (1832); and Wallich nowhere produces the binomial *Centrostachys aquatica.* Hence this must be dated from Moquin.

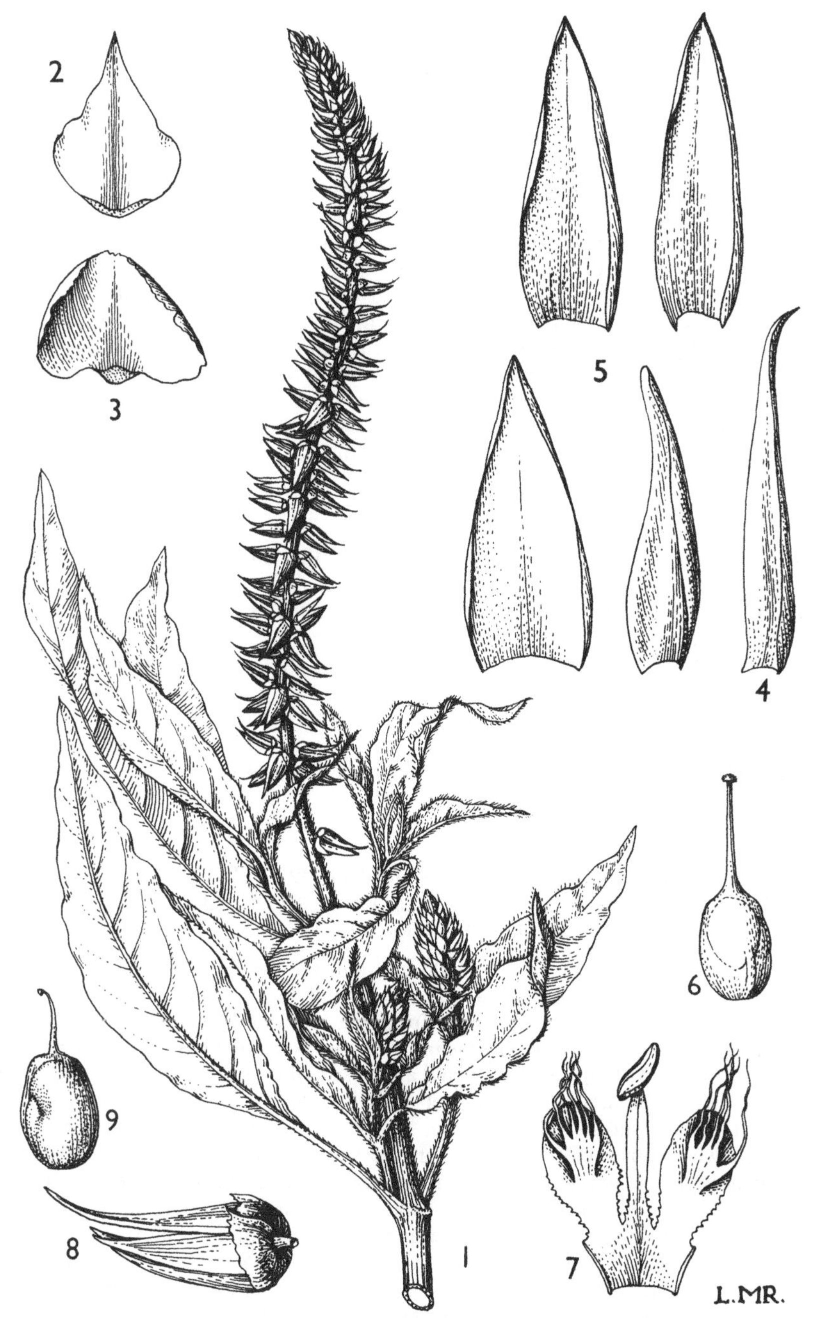

FIG. 24. *CENTROSTACHYS AQUATICA*—**1**, flowering branch, × 1; **2**, bract, × 6; **3**, bracteole, × 6; **4**, outermost tepal, × 6; **5**, inner tepals, × 6; **6**, gynoecium, × 6; **7**, part of androecium, × 6; **8**, fruiting perianth, × 4; **9**, capsule, ×4. 1–7, from *Greenway* 10146; 8, 9, from *Greenway & Kanuri* 15446. Drawn by Lura Ripley.

20. ACHYROPSIS

Hook.f. in G.P. 3: 36 (1880)

Annual or perennial herbs or low shrubs with entire opposite leaves which may be solitary or in fascicles. Inflorescence terminal on the stem and branches, subcapitate to spiciform, bracteate. Flowers solitary in the axils of the bracts, all ⚥, bibracteolate, small, modified sterile flowers absent. Bracts persistent, finally weakly deflexed or deflexed-ascending; bracteoles and perianth falling with the fruit; bracteoles closely appressed to the perianth. Perianth-segments 4–5, glabrous, firm, deeply concave and usually ± hooded at the apex, 1(–3)-nerved with the midrib ceasing below the muticous apex or occasionally excurrent in a minute mucro. Stamens 4–5; filaments delicate, shortly monadelphous at the base, alternating with narrowly triangular to quadrate or oblong pseudostaminodes with or without a dorsal scale; anthers bilocular. Ovary with a single pendulous ovule, glabrous; style very short to slender; stigma capitate. Fruit a thin-walled capsule, irregularly ruptured by the developing seed. Seed globose or slightly compressed; endosperm copious.

6 species in tropical and South Africa.

Tufts of long, white woolly hairs between the bracts and the perianth (Fig. 25/7) .... 1. *A. laniceps*
No tufts of long, white woolly hairs present between the bracts and the perianth:
  Leaves filiform, perfectly glabrous .... 2. *A. filifolia*
  Leaves neither filiform nor glabrous:
    Tepals delicate and hyaline, with a single slender midrib and no lateral nerves; plant annual.............. 3. *A. gracilis*
    Tepals firm, with 2–6 fine lateral nerves in addition to the midrib; plant a low shrub .... 4. *A. fruticulosa*

1. **A. laniceps** *C.B. Cl.* in F.T.A. 6(1): 66 (1909); F.D.O.-A. 2: 240 (1932); Cavaco in Mém. Mus. Nat. Hist. Nat. Paris, sér. B, 13: 152 (1962). Type: Malawi, Chitipa [Fort Hill], *Whyte* (K, holo.!)

Annual with slender tap-root, 15–100 cm. tall, the smaller forms much reduced with an unbranched stem and a single terminal inflorescence, the larger with numerous slender branches diverging from the stem at about 45°, each branch bearing 1(–2) pairs of leaves below the inflorescence; stem and branches tetragonous, striate, furnished with upwardly directed appressed white hairs. Leaves of main stem linear to linear-oblanceolate, (1–)2.2–6(–9) × (0.1–)0.2–0.7(–0.9) cm., narrowed to each end, not or indistinctly petiolate, mucronate at the apex, green and thinly furnished with appressed white hairs on the upper surface, generally whitish and more densely pilose below; upper stem and branch leaves much reduced. Peduncles slender, very variable in length (that of the terminal inflorescence up to ± 12 cm.), indumentum similar to that of the stem; inflorescence at first short and conical, becoming cylindrical, woolly, white or pink-tipped, 1–2.5 × 0.4–0.7 cm., nodding; axis white lanate; bracts lanceolate, 2 mm., aristate with the excurrent midrib, hyaline, glabrous or ± long-ciliate, persistent, with many long white hairs between the bracts and the flowers; bracteoles closely appressed to the perianth, subrotund, ± 0.6–0.75 mm., glabrous, falling with the perianth. Tepals all very concave, hooded above, similar, 1.75–2 mm., with a single very slender midrib which is not excurrent in a distinct mucro, glabrous, broadly white-margined with a green central vitta, occasionally flushed with pink. Filaments ± 1–1.5 mm., subulate; anthers subrotund; pseudostaminodes quadrate or rounded, with a narrow, truncate or rounded ventral flap and a narrowly oblong to linear dorsal scale which is fimbriate at the apex and equals or exceeds the filaments. Ovary obpyriform; style very short, to ± 0.4 mm. Capsule subglobose, ± 1.25 mm., delicate below, with a small firm apex. Seed filling the capsule, globose, ± 1.25 mm., yellow-brown, smooth and somewhat shining, feebly reticulate. Fig. 25/6–8.

TANZANIA. Mbulu District: Pienaars Heights, *Wright* 5!; Dodoma District: E. of Itigi Station, Apr. 1964, *Greenway & Polhill* 11451!; Songea District: below Matagoro Hills just S. of Songea, May 1956, *Milne-Redhead & Taylor* 9870!

DISTR. T2, 4, 5, 7, 8; Zaire, Burundi, Malawi, Zambia

HAB. In open *Brachystegia* woodland or bush, in sandy or sandy-loamy places, generally among grass and often where water has lain, also in pockets of soil over rock; 900–1820 m.

SYN. *Achyropsis robynsii* Schinz in Viert. Nat. Ges. Zürich 76: 143 (1931); Hauman in F.C.B. 2: 41 (1951). Type: Zaire, Kipushé, *Robyns* 1812 (BR, holo.!)
A. laniceps C.B.Cl. forma *robynsii* (Schinz) Cavaco in Mem. Mus. Nat. Hist. Nat. Paris, sér. B, 13: 153 (1962)

2. **A. filifolia** *C.C. Townsend* in K.B. 34: 431 (1979). Type: Tanzania, Mbeya District, Ruaha National Park, Magangwe Hill, *Bjφrnstad* 2687 (O, holo.!, K, iso.!)

Annual with slender taproot, 20–30 cm. tall, simple or with a few long divergent branches from near the base, quite glabrous except for the lanate inflorescence-axis; stem and branches very slender, terete, striate. Leaves all filiform, 15–55 × 0.5–1 mm., mucronate at the apex, margins and back of nerve slightly scabrid. Peduncles long and very slender, 2–13 cm., striate. Inflorescence at first capitate, finally shortly cylindrical, 0.6–1 × 0.4 cm.; bracts deltoid-ovate, persistent, patent, ± 1.25 mm., glabrous, rather blunt, with a greenish central vitta and broad hyaline margins, the slender nerve reaching to the apex but scarcely excurrent in a mucro; bracteoles similar, closely appressed to the perianth. Perianth somewhat pyriform, narrowed about the middle; outer 2 tepals ovate, very concave, hooded above, ± 2 × 1 mm., feebly 3-nerved with the midrib almost attaining the apex but not excurrent in a distinct mucro, green-vittate centrally with a hyaline border widening from the middle downwards; middle tepal slightly narrower, otherwise similar; inner 2 lanceolate-oblong, abruptly narrowed to the tip, ± 0.5 mm. wide, lateral nerves not obvious, hyaline margin wider. Filaments very slender, subulate, ± 1.75 mm.; pseudostaminodes ± 0.75 mm., oblong, expanded and fimbriate at the apex. Ovary obpyriform, ± 1 mm.; style slender, ± 0.75 mm. Capsule and ripe seed not known.

TANZANIA. Mbeya District: Ruaha National Park, summit of the NW.-most hillock of Magangwe Hill, Mar. 1973, *Bjφrnstad* 2687!

DISTR. T 7; known only from the type gathering

HAB. *Brachystegia* woodland; 1540 m.

NOTE. Very similar in aspect to *Centemopsis longipedunculata*, but with blunter, more concave, cucullate tepals and quite different pollen.

3. **A. gracilis** *C.C. Townsend* in K.B. 34: 432 (1979). Type: Tanzania, Lushoto/Handeni Districts, Handeni road (? from Korogwe), *Faulkner* 4235 (K, holo.!, EA, iso.!)

Spreading annual (?) with a slender rootstock, 30–40 cm., rooting at some of the lower nodes, sparingly branched in the lower part of the stem with branches divergent at 45° and then ascending; stem and branches tetragonous and striate, moderately furnished with rather long flexuose hairs. Leaves elliptic-ovate, 2–4 × 0.7–1.3 cm., cuneate into a short (2–4 mm.) petiole below, shortly pointed and subacute at the apex, paler on the lower surface, both surfaces moderately furnished with fine whitish subappressed multicellular hairs. Peduncles 1–4.5 cm., indumentum

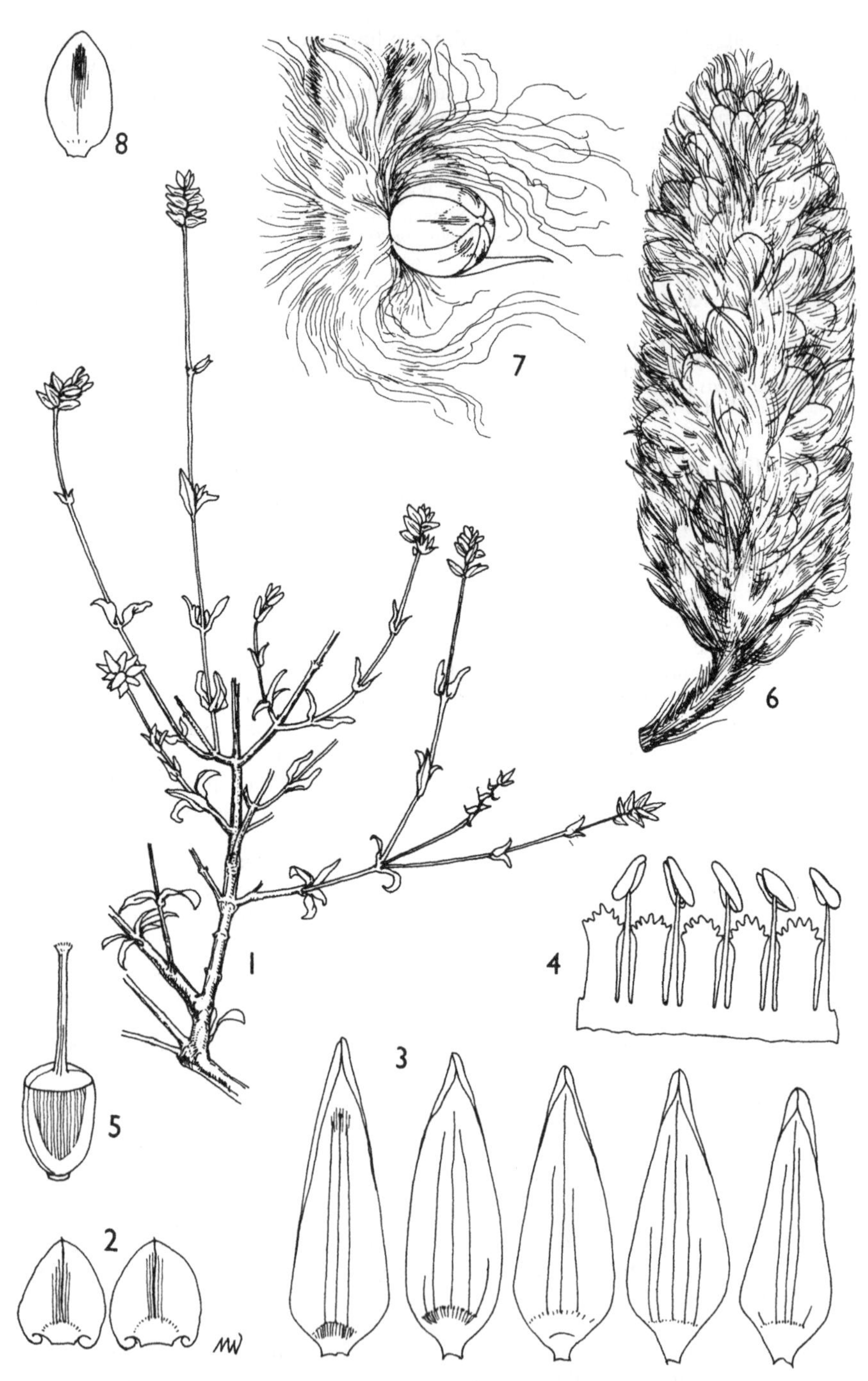

FIG. 25. *ACHYROPSIS FRUTICULOSA*—**1**, flowering branch, × $\frac{2}{3}$; **2**, bracteoles, × 10; **3**, tepals, × 10; **4**, part of androecium, × 10, **5**, gynoecium, × 10. *A. LANICEPS*—**6**, inflorescence, × 4; **7**, single flower on inflorescence-axis, × 7; **8**; bracteole, × 7. 1-5, from *Glover* 887; 6-8, from *Greenway* 342. Drawn by Mary Millar Watt.

similar to that of the stem; inflorescence at first conical-capitate, becoming elongate with the flowers increasingly distant below, 2–6 × 0.6 cm.; axis white lanate; bracts lanceolate-ovate, 2 mm., aristate with the excurrent midrib, finely ciliate, persistent; bracteoles suborbicular-ovate, ± 1.25 mm., finely ciliate, hyaline, the midrib excurrent in a short fine but distinct mucro. Tepals elliptic, reducing in width from the outer to the inner, membranous, whitish save for a narrow central green vitta, glabrous, 3 mm., the slender midrib excurrent in a very short mucro. Stamens ± 2 mm.; filaments filiform; pseudostaminodes oblong with a truncate ventral scale, the dorsal scale long-fimbriate with the cilia subequalling the filaments. Ovary obpyriform; style slender, ± 1 mm. Capsule subglobose, 1.5 mm. Seed chestnut, filling the capsule, apiculate beneath the style-base, shining, feebly reticulate.

TANZANIA. Lushoto / Handeni Districts: 16 km. on Handeni road(? from Korogwe), June 1969, *Faulkner* 4235!
DISTR. **T** 3; Mozambique
HAB. On floor of forest, semi-shaded; 152 m.

4. **A. fruticulosa** *C.B.Cl.* in F.T.A. 6(1): 66 (1909). Type: Kenya, Kiambu District, Kukui, *Kassner* 1011 (K, holo.!)

Low shrub 17–120 cm. tall, much branched, the older parts with a woody frequently gnarled growth and yellowish grey cortex, terete, glabrescent; younger branches divergent, passing to tetragonous, striate, ± densely furnished with greyish upwardly directed appressed hairs. Leaves lanceolate to narrowly oblong-elliptic, 4–40 × 1.5–11 mm. (frequently with sterile axillary shoots and thus appearing fasciculate), subacute to acute at the apex, sessile or very shortly petiolate at the base, densely furnished with appressed white hairs on the lower surface but usually greener and more thinly hairy above, margins frequently somewhat undulate or revolute. Peduncles rather stout, 1.5–6 cm., striate, indumentum similar to that of the younger branches, inflorescence cylindrical, 1.4–4(–7) × 0.6–0.8 cm., grey-green; axis lanate; bracts lanceolate- to deltoid-ovate, ± 2–2.5 mm., shortly pilose, scarious, ± stramineous, the brownish midrib excurrent in a short arista, persistent; bracteoles closely appressed to the perianth, broadly deltoid- to suborbicular-ovate, 1.25–1.5 mm., hyaline-membranous, brownish at the centre base, glabrous or ciliate-margined, the brownish midrib reaching the apex but scarcely excurrent into a mucro, falling with the perianth. Tepals very concave, greenish with a hyaline margin, blunt with the fine midrib not excurrent in a mucro, the outer with 1–3 pairs of fine lateral nerves and broadly lanceolate-ovate, 3–4 mm., the inner gradually somewhat shorter and narrower, usually (but not always) with fewer nerves; all glabrous. Filaments subulate, firm, 1.5–2 mm., anthers subrotund; pseudostaminodes narrowly triangular and toothed ± all round to narrowly oblong with a fringed apex, usually slightly shorter than the filaments. Ovary obpyriform to shortly ovoid-oblong; style slender, 1.75–2 mm. In spite of much material with perianths fallen and on the point of falling being available, no ripe capsules or seeds have seen; the most mature capsules are oblong-ovoid, ± 1.25 mm., delicate below with a firm convex apex. Fig. 25/1–5.

KENYA. N. Nyeri District: Coles Mill, Jan. 1922, *R.E. & Th. C.E. Fries* 1086!; Nairobi District: 27 km. E. of Nairobi, Nairobi R. valley, Sept. 1951, *Bogdan* 3265!; Masai District: 24–32 km. Narok to Mara R., June 1956, *Verdcourt* 1505!
TANZANIA. Masai District: Serengeti National Park, 1970, *Krevlen* 53!; Moshi District: Shira, Mar. 1914, *Peter* 2701! & Ngare Nairobi, July 1934, *Staples* 372!
DISTR. **K** 1, 3, 4, 6, 7; **T** 2; not known elsewhere
HAB. Almost exclusively on dry grassland with herbs and low bushes, sometimes with *Acacia*, once recorded from a forest edge and once from a damp place in a field, only soil recorded being black cotton; 600–1880 m.

SYN. *Achyranthes fischeri* Fries in N.B.G.B. 9: 319 (1925); F.D.O.-A. 2: 240 (1932), *nomen*
Pandiaka fischeri sensu Peter in F.D.O.-A. 2: 242 (1938), *nomen*
Achyropsis greenwayi Suesseng. in K.B. 4: 475 (1950); U.K.W.F.: 137 (1974). Type:

Tanzania, Masai District, Engare Rongai [Ngari Rongi] Plain, *Greenway* 6740 (K, holo.!, EA, iso.)

NOTE. A note on one sheet from Loita Plains (*Glover, Gwynne & Samuel* 750) states that this plant is grazed by all domestic stock.

21. **PANDIAKA**

(Moq.) Hook.f. in G.P. 3: 35 (1880)

Achyranthes sect. *Pandiaka* Moq. in DC., Prodr. 13(2): 310 (1849)

Annual or perennial herbs with opposite, petiolate or sessile, entire leaves; rootstock slender to tuberous. Inflorescence spicate to capitate, elongating later or not, flowers spreading or in one species becoming deflexed after flowering. Flowers solitary in the bracts, ⚥. Bracteoles 2, the midrib excurrent in a short, glabrous and pungent or longer, flexuose and pilose arista, lamina firm, chartaceous or horny, ± half the length of the perianth or more. Perianth-segments 5, feebly to strongly 1–3(–5)-nerved, lanceolate to narrowly oblong, usually hairy (rarely glabrous), ± mucronate to aristate with the excurrent nerve. Stamens 5; filaments filiform, monadelphous below, alternating with mostly quadrate pseudostaminodes, these glabrous to pilose, simple or with a highly developed, fimbriate dorsal scale; anthers bilocular. Ovary with a solitary pendulous ovule, fruiting wall very thin; apex at maturity firm, with a transverse crest passing through the base of the style, often forming a hump at each end where it meets the periphery, rarely without such a crest but with a circumferential rim or almost flat; style slender; stigma small, truncate-capitate. Perianth falling with the ripe capsule, accompanied by the bracteoles or not; bracts persistent, spreading or rarely deflexed.

About 12 species, confined to tropical Africa.

Very close to *Achyranthes* indeed on technical characters, but the species of the latter genus have a characteristic appearance which makes them recognisable on sight.
Cavaco in Mém. Mus. Nat. Hist. Nat. Paris, sér. B. 13: 130 (1962) subdivides the genus as follows.
Subgen. **Pandiaka.** Leaves sessile; perianth not deflexing in fruit. Species 1–4.
Subgen. **Achyranthopsis** *Cavaco.* Leaves petiolate; perianth somewhat deflexing in fruit. Species 5. [Cavaco included *Achyranthes fasciculata* in this subgenus – see Townsend in K.B. 34: 423 (1979)].

Bracteoles and perianth densely white lanuginose; fruiting perianth sharply deflexed .... 5. *P. lanuginosa*
Bracteoles and perianth pilose with straight hairs, or the bracteoles ± glabrous; fruiting perianth not deflexed:
 Ovary (Fig. 26/6) and fruit with a firm circumferential rim, concave at maturity; inflorescences subtended by an "involucre" of 2–4 linear ± parallel-sided leaves ..... 1. *P. angustifolia*
 Ovary and fruit with a transverse keel or faint line but with no circumferential rim; inflorescences subtended by 2–4 ovate to lanceolate leaves, or at least some without any involucre:
  Leaves expanded and ± auriculate-amplexicaul at the base; inflorescences spiky, with longly and sharply aristate tepals .... 3. *P. rubro-lutea*
  Leaves sessile but not auriculate-amplexicaul at the base; inflorescences not spiky, the tepals rigidly pointed but not longly and sharply aristate:
   Inflorescences sessile on 2–4 broad-based leaves, the bracteoles subequalling the perianth; annual .... 2. *P. involucrata*
   Inflorescences not involucrate on 2–4 broad-based leaves,

the bracteoles much shorter than the perianth; perennial .... 4. *P. welwitschii*

1. **P. angustifolia** *(Vahl) Hepper* in K.B. 25: 189 (1971). Type: probably Ghana, and probably collected by Isert, but labelled as "dedit Thonning" (S, holo., photo.!)

Annual herb, usually considerably divaricately branched, 0.2–0.8(–1.2) m.; stem and branches strongly ridged, often ± red, moderately to densely pilose with upwardly directed ± appressed hairs. Leaves linear, 15–80 × 1–11(–14) mm., narrowed to both ends, apex mucronate, thinly to moderately pilose on the upper surface, moderately to densely so on the lower, hairs appressed. Inflorescence white or frequently pink-tinged especially about the tip, subglobose to cylindrical, (1–)1.6–5 cm. long and 1.2–1.8 cm. wide, sessile at the ends of stem and branches, on an involucre of 2–4 leaves up to 6.5 cm. long. Bracts broadly ovate or elliptic-ovate, 3–5.5 mm., white-membranous, glabrous or finely ciliate-margined, the midrib excurrent in a pale sharp ± 1 mm. arista. Bracteoles ovate, white-membranous, sparingly to considerably furnished with long white hairs along the dorsal surface of the midrib, (3.5–)5–8 mm., the midrib excurrent into a ± 1–2 mm. ± recurving arista. Flowers sessile. Tepals firm, opaque, lanceolate, narrowly hyaline-margined, usually greenish centrally; outer 2 at first 4.5–6 mm., equalling or slightly exceeding the bracteoles, lengthening to 7–8 mm. and clearly exceeding the bracteoles in fruit, ± densely furnished with long variably spreading white hairs, 5–7-nerved in transmitted light, the tips erect or ± spreading, pale or brownish, awn-like; inner 2 slightly shorter and narrower, obscurely 3-nerved, glabrous, tips usually erect; middle tepal intermediate, sparingly pilose; all indurate, white and shining at the base in fruit, fused below the ovary by incurved appendages, with a central basal cavity set over a cone-shaped protuberance by which the bracteoles are joined. Fruiting perianth first falling from the bracteoles, and later the bracteoles from the bract. Stamens 2.5–3 mm., pseudostaminodes ± 0.75–1 mm., oblong, truncate-excavate with an incurved lobule at the tip, dorsal scale minute. Ovary oblong-ovoid, ±0.75 mm., with a firm scutellate apex; style slender, 2–3 mm. Capsule oblong-ovoid, 2 mm.; apex cup-shaped with the style set in the base of the concavity, basal part hyaline and delicate below the firm circumferential rim. Seed oblong-ellipsoid, 2 mm., brown, shining. Fig. 26.

UGANDA. W. Nile District: Koboko, Mar. 1938, *Hazel* 430! & Terego, July 1938, *Hazel* 634!; Bunyoro, Oct. 1961, *Turner* 85!
TANZANIA. Kigoma District: Kabogo Mts., May 1962, *Azuma* 518!
DISTR. U 1, 2; T 4; W. tropical Africa from Mauritania S. to Nigeria, Cameroun, Burundi, Angola and Malawi, E. to Chad and Sudan
HAB. Open grassland, roadside, *Combretum* savanna, open *Brachystegia* woodland; 750–1100 m.

SYN. *Gomphrena angustifolia* Vahl, Symb. Bot. 3: 45 (1794)
Achyranthes heudelotii Moq. in DC., Prodr. 13(2): 310 (1849). Type: Senegal, Bakel, *Heudelot* 280 (P, holo., photo.!)
A. angustifolia Benth. in Hook., Niger Fl.: 492 (1849); P.O.A.C: 1174 (1895). Type: Nigeria, R. Niger [Quorra], *Vogel* 98 (K, holo.!)
Pandiaka heudelotii (Moq.) Hiern, Cat. Afr. Pl. Welw. 1: 894 (1900); F.T.A. 6(1): 68 (1909); F.D.O.-A. 2: 241 (1932); F.P.S. 1: 120 (1950); Hauman in F.C.B. 2: 45 (1951); Cavaco in Mém. Mus. Nat. Hist. Nat. Paris, sér. B, 13: 135 (1962)
Achyranthes angustifolia (Vahl) Lopr. in E.J. 30: 107 (1901) & in Malpighia 14: 434 (1901)
A. benthamii Lopr. in E.J. 30: 108 (1901) & in Malpighia 14: 435 (1901). Type: as for *A. angustifolia* Benth.
Pandiaka benthamii (Lopr.) Schinz in E. & P. Pf., ed. 2, 16C: 64 (1934); Hauman in F.C.B. 2: 45 (1951)
P. heudelotii (Moq.) Hiern var. *spicata* Suesseng. in B.J.B.B. 15: 65 (1938). Type: Ghana, *Irvine* 542 (K, lecto.!)
P. heudelotii (Moq.) Hiern var. *subglobosa* Suesseng. in B.J.B.B. 15: 66 (1938), cum forma *rubella*. Types: Sudan, Djur, *Schweinfurth* 2309, var. (K, lecto.!); Nigeria, *Lely* P641, forma (K, holo.!)

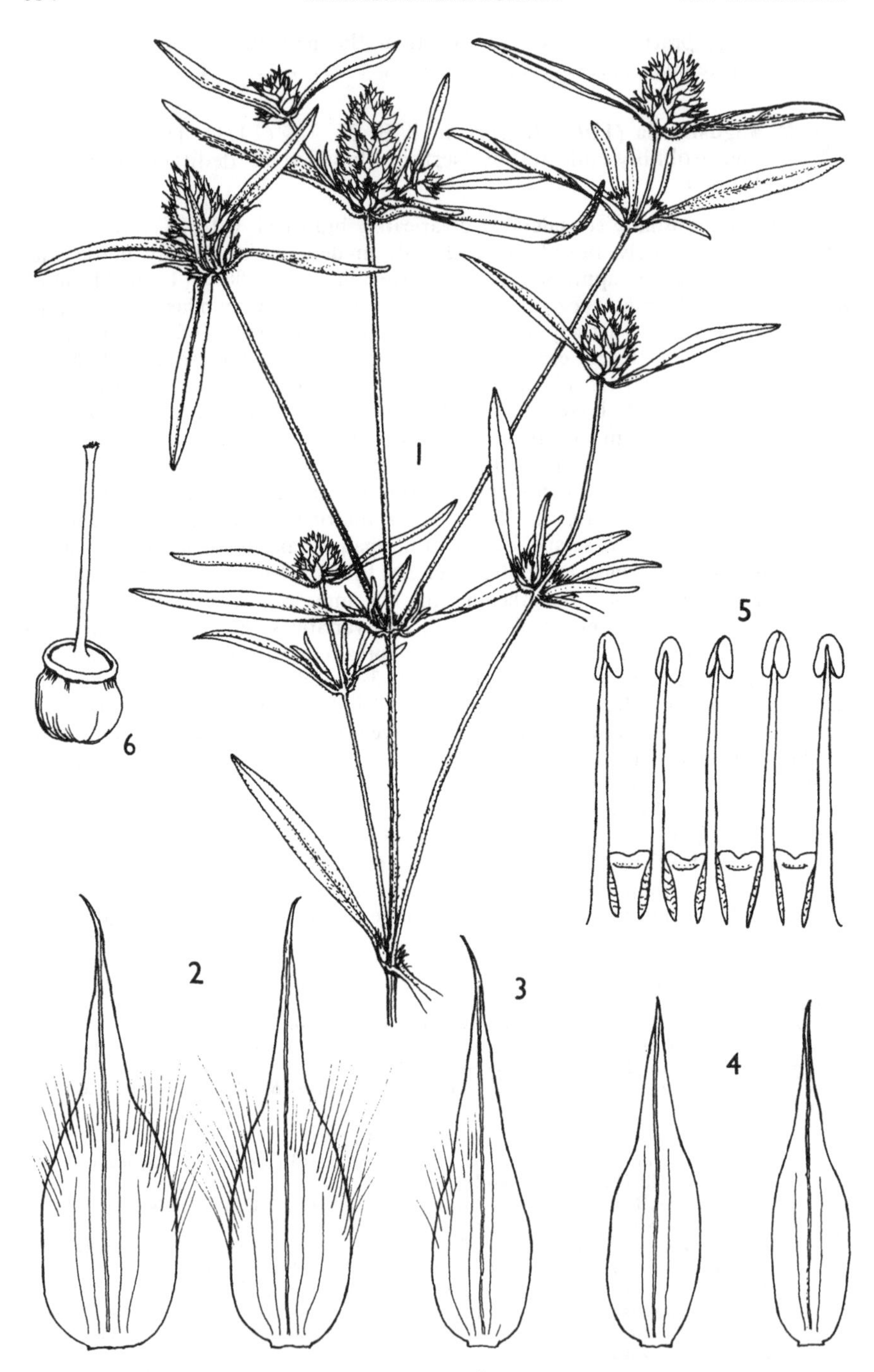

FIG. 26. *PANDIAKA ANGUSTIFOLIA*—**1**, flowering branch, × $\frac{2}{3}$; **2**, outer tepals, × 9; **3**, intermediate tepal, × 9; **4**, inner tepals, × 9; **5**, part of androecium, × 9; **6**, gynoecium, × 9. All from *Scott* 11794. Drawn by Mary Millar Watt.

2. P. involucrata *(Moq.) B.D. Jackson* in Index Kewensis: 409 (1894); F.T.A. 6(1): 67 (1909); F.W.T.A, ed. 2, 1: 151 (1954); Cavaco in Mém. Mus. Nat. Hist. Nat. Paris, sér. B, 13: 133 (1962). Type: N. Nigeria, Pandiaki, *Ansell* 9/17/41 (K, holo.!)

Annual herb, simple in very reduced plants, occasionally branched from the base with decumbent branches, but usually erect with many long, slender, erect branches, (0.25–)0.45–1.2(–2) m. tall; stems frequently red, terete below, ± quadrangular above, striate, clad (as are the branches) with an indumentum of ± dense, upwardly directed to subpatent whitish hairs. Leaves lanceolate-oblong to oblong-elliptic, those of the main stems 2–10 × 0.9–3 cm., ± abruptly contracted at the base (upper leaves sessile, the lower usually very shortly petiolate), obtuse-apiculate to subacute at the apex, thinly to moderately furnished with upwardly-directed or subpatent hairs, the upper surface usually darker. Inflorescences subglobose to cylindrical, (1–)1.5–3.5(–5) × (1.25–)1.5–1.75 cm., the flowers very dense and concealing the white-pilose axis, sessile above 2–4 lanceolate to ovate, short leaves; commonly in groups of 3, a terminal with two laterals on widely spreading branches. Bracts ovate-acuminate, ± 5–8 mm., membranous with a narrow or wider greenish central band, very shortly pilose along the midrib only to more generally long-pilose, the midrib excurrent in a distinct, sharp arista. Bracteoles lanceolate-ovate, 6–9 mm., densely long-pilose at least along the midrib dorsally, the midrib excurrent in a sharp mucro, the tip of one bracteole (and not rarely the entire flower) ± curved. Tepals narrowly lanceolate, the outer 2 densely appressed-pilose, narrowly pale-margined with a firm green centre, 6–8.25 mm.; inner 2 ± 0.5–1 mm. shorter, less pilose, 3 nerves sometimes visible in transmitted light only; middle tepal intermediate; all with a rigid, spine-like tip. Stamens 2.5–5 mm., the filaments floccose-hairy near the base; pseudostaminodes flabellate, 0.75–1 mm., the margins ± floccose, dorsal scale absent in all material examined. Ovary quadrate-obpyriform, slightly compressed, with a faint apical keel; style slender, (1–)2.5–3.25 mm. Capsule turbinate ("drum-shaped"), ± 2.5 mm., with a broad, flat, firm apex; seeds oblong-ovoid, 2–2.25 mm., brown, shining, faintly reticulate.

TANZANIA. Ufipa District: escarpment above Kasanga, Mar. 1959, *Richards* 12297!
DISTR. T4; widespread in W. Africa eastwards to Kordofan, apparently absent from Zaire, southwards to the Victoria Falls region and central Malawi
HAB. Roadside in rocky ground and gritty soil; 900 m.

SYN. *Achyranthes involucrata* Moq. in DC., Prodr. 13(2): 310 (1849)
Pandiaka involucrata (Moq.) B.D. Jackson var. *megastachya* Suesseng. in F.R. 44: 46 (1938). Type: N. Nigeria, Naraguta, *Lely* 239 (K, holo.!)

NOTE. The only East African gathering known, that cited above, has large inflorescences up to 6 cm. long, and is a good match for the type of "var. *megastachya*".
As Moquin realised when he described *Achyranthes involucrata*, the earlier name. *Achyranthes nodosa* Vahl ex Schum. virtually certainly refers to the present plant; but the absence of a type specimen and one or two discrepancies in the description of *A. nodosa* (particularly as to its habit) make good enough reason not to displace a well-known name.

3. **P. rubro-lutea** *(Lopr.) C.C. Townsend* in K.B. 34: 428 (1979). Type: Zaire, Shaba, Pueto [M'Pueto], *Descamps* (Z, lecto.!)

Annual herb, usually considerably branched (the branches divaricate-ascending) but simple in poorly developed specimens, 0.2–0.9 m.; stem and branches ridged, often reddish, moderately to rather densely pilose with upwardly directed ± appressed hairs. Leaves broadly obovate to linear, 1.5–8.5 × 0.2–3.5 cm., thinly to moderately pilose, sessile and in the broader-leaved forms commonly constricted ± ⅓ of the way up and then expanded to an auriculate base, apex blunt to subacute, obscurely mucronate. Inflorescence green or the tips of bracts, bracteoles and tepals frequently pink to carmine, conical when young but finally subglobose to cylindrical,

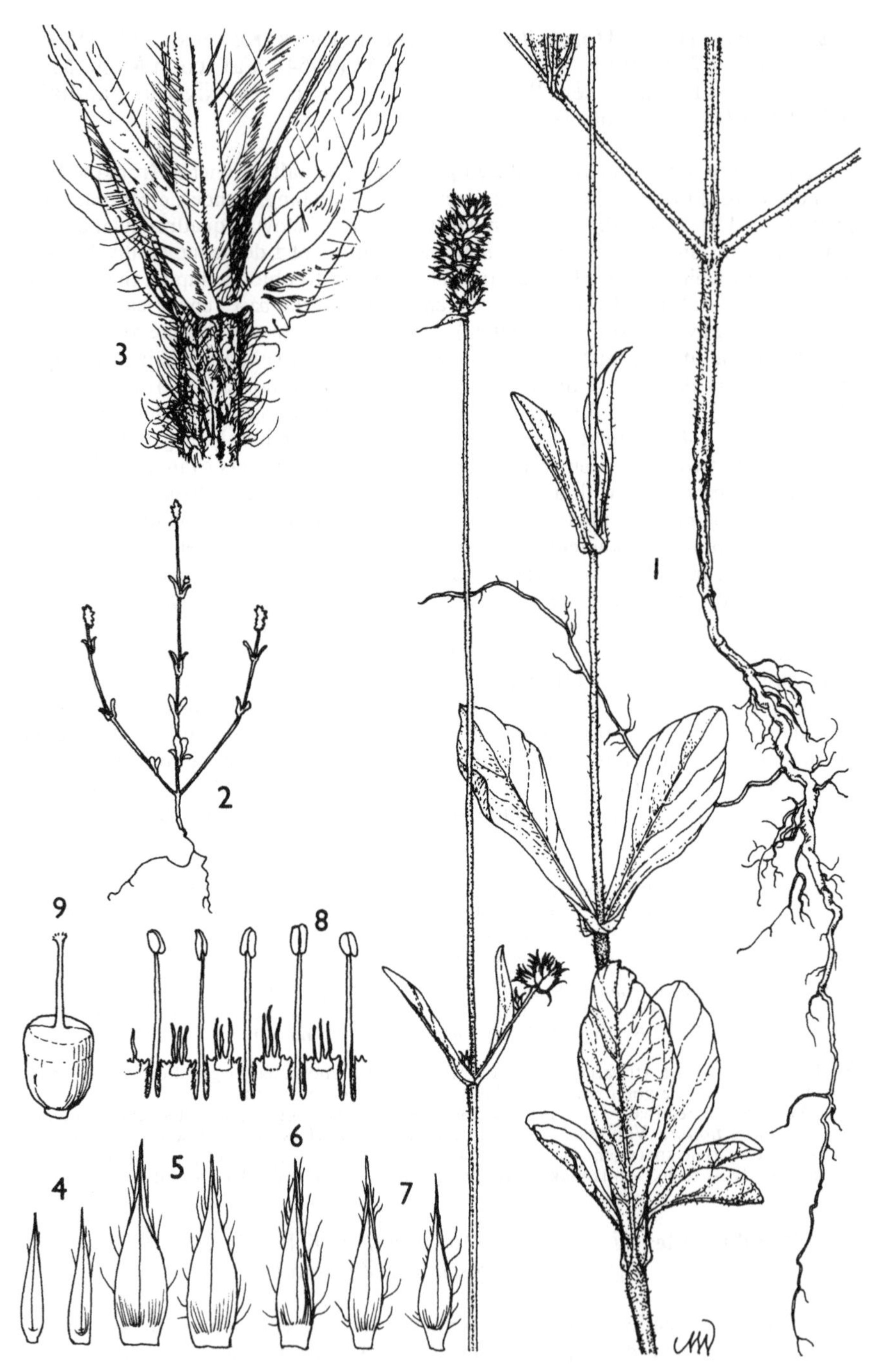

FIG. 27. *PANDIAKA RUBRO-LUTEA*—**1**, flowering plant, × ⅔ 3; **2**, habit, much reduced; **3**, node showing leaf-bases, × 6; **4**, bracteoles, × 5.5; **5**, outer tepals, × 5.5; **6**, intermediate tepal, × 5.5; **7**, inner tepals, × 5.5; **8**, part of androecium, × 8; **9**, ovary, × 10. All from *Milne-Redhead & Taylor* 9422A. Drawn by Mary Millar Watt.

1.2–6.5 × 1.2–1.4 cm., very dense, on a short to long (up to 11 cm.) peduncle or more rarely sessile through branch-condensation and then often with short lateral inflorescences at the base; axis villous. Bracts ovate-lanceolate, 3–5 mm. white-membranous, sparingly pilose along the central dorsal surface or glabrous, gradually tapering to the acute 1–1.5 mm. arista formed by the excurrent midrib. Bracteoles narrower, lanceolate, 2.75–5 mm., more densely pilose, the arista 1.75–2.5 mm. Flowers truncate at the base, where they are indurate, fused to the bracteoles and attached to the inflorescence-axis by a knob-like stalk. Tepals lanceolate-subulate, greenish centrally, firm and opaque except for the narrowly hyaline margin, the stout midrib excurrent in a very sharp pale to reddish, 1–2 mm. arista, erect or divergent at the tips; 2 outer 5.5–7 mm., ± densely furnished with upwardly directed ± spreading white hairs, with 5 nerves which are often white and obvious near the base but obscured above; 2 inner ± 1 mm. shorter, narrower, 3-nerved, hairs mostly confined to the central and upper parts; middle tepal intermediate. Perianth and bracteoles falling in one unit in fruit. Stamens 2–3 mm.; pseudostaminodes ± 0.75–1 mm., oblong, truncate with an incurved lobule at the apex, dorsal scale filiform to considerably fimbriate. Ovary roundish, ±1 mm., with a firm, transversely ridged apex; style slender, 1–1.5 mm. Capsule oblong-ovoid, 2–2.25 mm., the firm flat apex with a transverse crest on each side of the style. Seed oblong-ellipsoid, ±2 mm., brown, shining. Fig. 27.

TANZANIA. Mwanza District: Mbarika, Apr. 1953, *Tanner* 1387!; Iringa District: Mbagi–Mdonya R. Track, Apr. 1970, *Greenway & Kanuri* 14399!; Songea District: 1 km. S. of Gumbiro, May 1956, *Milne-Redhead & Taylor* 9422A!

DISTR. T1, 4, 5, 7, 8; Zaire (Shaba), Rwanda, Burundi, Angola, Zambia, Zimbabwe, Malawi

HAB. Sides of roads and tracks in light shade, dry open woodland, light scrub, rock crevices, usually on sandy soil; 240–1620 m.

SYN. *Achyranthes rubro-lutea* Lopr. in E.J. 27: 57 (1899); F.T.A. 6(1): 65 (1909); F.C.B. 2: 55 (1951)

Pandiaka andongensis Hiern, Cat. Afr. Pl. Welw. 1: 895 (1900); F.T.A. 6(1): 70 (1909); Hauman in F.C.B. 2: 43 (1951); Cavaco in Mém. Mus. Nat. Hist. Nat. Paris, sér. B, 13: 144 (1962). Type: Angola, Pungo Andongo, *Welwitsch* 8567 (BM, K, iso.!)

P. lindiensis Suesseng. & Beyerle in F.R. 44: 46 (1938); Cavaco in Mém. Mus. Nat. Hist. Nat. Paris, sér. B, 13: 143 (1962). Type: Tanzania, Lindi District, Lake Lutamba, *Schlieben* 6083 (K, iso.!)

P. aristata Suesseng. in B.J.B.B. 15: 64 (1938); Cavaco in Mém. Mus. Nat. Hist. Nat. Paris, sér. B, 13: 145 (1962). Type: Tanzania, Iringa, *Lynes* Ih.215e (K, holo.!)

P. andongensis Hiern var. *gracilis* Suesseng. in Mitt. Bot. Staats., München 1: 341 (1953). Type: Tanzania, Mwanza District, Mbarika, Mwanghanda, *Tanner* 767 (K, holo.!)

4. **P. welwitschii** *(Schinz) Hiern*, Cat. Afr. Pl. Welw. 1: 894 (1900); F.T.A. 6(1): 69 (1909); F.D.O.-A. 2: 241 (1932); Cavaco, in Mém. Mus. Nat. Hist. Nat. Paris, sér. B, 13: 141 (1962). Type: Angola, Huila, *Welwitsch* 6488 (BM, K, iso.!)

Perennial herb, (15–)50–120 cm., taller forms much branched with the lower branches increasingly widely divaricate; stem and branches ± densely furnished with whitish to yellowish spreading to upwardly ascending hairs, quadrangular, the older parts ± glabrescent and terete. Leaves usually obovate and rounded-apiculate, sometimes elliptic and subacute, broadly tapering below, those of the main stem and branches 1.2–8×0.6–3.5 cm., moderately appressed pilose on both surfaces with hairs usually of varying lengths, darker green above and paler beneath. Inflorescences terminal on the stem and branches, sometimes ± clustered at the top of the stem by branch reduction, pinkish, conical when young but finally cylindrical, 3–10 × 1.25–1.5 cm., very dense, on a 0.5–4 cm. densely tomentose peduncle. Bracts lanceolate-ovate, 4–6 mm., membranous with an obscurely to distinctly green centre, ± densely pilose centrally, diminishing to glabrous at the margins, tapering to the short acute arista formed by the excurrent midrib. Bracteoles similar or slightly narrower, 3–5 mm., more densely pilose, longer (± 0.5–1 mm.) aristate. Flowers truncate at the base and indurate, attached to the inflorescence-axis by a white annular callus. Tepals narrowly lanceolate; outer 2 greenish and ± densely pilose centrally, 5–8 mm., with 3(–5) white nerves, the 1(–2) outer pairs much shorter than the midrib, which is excurrent in a short arista; inner 3 similar but slightly shorter and progressively somewhat narrower and less pilose, 3-nerved or the innermost occasionally with only the midrib showing; all usually slightly upwardly curving near the tip. Perianth and bracteoles falling together in fruit. Stamens 2.5–5 mm.; pseudostaminodes 1–1.5 mm., oblong to flabellate, fimbriate (or dentate only at the apex with the margins fimbriate); dorsal scale broader and fimbriate to subulate-dentate. Ovary obpyriform, ± 1.5 mm.; style slender, 2.75–4.25 mm. Capsule oblong-ovoid, ± 2 mm., the firm flat apex with a transverse crest on each side of the style. Seed oblong-ellipsoid, ± 1.75 mm., brown, shining.

TANZANIA. Biharamulo District: Lusahanga, 15 Oct. 1960, *Tanner* 5588A; Mwanza District: Pasiansi, May 1969, *Mbano* in *DSM.* 775!; Mpanda District: Kabungu, Aug. 1948, *Semsei* 90!
DISTR. T 1, 4; Cameroun, Zaire, Sudan, Angola, Zambia
HAB. Deciduous woodland, grassland, rocky hillsides and riversides, on sandy soil; 1125-1820 m.

SYN *Achyranthes welwitschii* Schinz in E.J. 21: 187 (1895) & in P.O.A. C: 174 (1895)
A. schweinfurthii Schinz in Bull. Herb. Boiss. 4: 421 (1896). Type: Sudan, Gudju, *Schweinfurth* ser. 3, 66 (Z, holo.!)
Psilotrichum debile Bak. in K.B. 1897: 279 (1897). Type: Angola, Huila, *Welwitsch* 6570 (K, holo.!)
Pandiaka debilis (Bak.) Hiern, Cat. Afr. Pl. Welw. 1: 894 (1900); F.T.A. 6(1): 69 (1909)
P. schweinfurthii (Schinz) C.B. Cl. in F.T.A. 6(1): 69 (1909); F.D.O.-A. 2: 241 (1932); F.P.S. 1: 120 (1950); Hauman in F.C.B. 2: 48 (1951); Cavaco in Mém. Mus. Nat. Hist. Nat. Paris, sér. B, 13: 141 (1962)
P. welwitschii (Schinz) Hiern var. *debilis* (Bak.) Suesseng. in F.R. 44: 47 (1938)
Achyranthes kassneri sensu Peter in F.D.O.-A. 2: 240 (1938), *nomen.*
Pandiaka kassneri Suesseng. in B.J.B.B. 15: 67 (1938); Hauman in F.C.B. 2: 46 (1951); Cavaco in Mém. Mus. Nat. Hist. Nat. Paris. sér. B, 13: 145 (1962). Type: Zaire, Shaba, Kasanga, *Kassner* 2665 (K, Z, iso.!)

5. **P. lanuginosa** *(Schinz) Schinz* in E. & P. Pf., ed. 2, 16C: 64 (1934); Cavaco in Mém. Mus. Nat. Hist. Nat. Paris, sér. B, 13: 147 (1962). Type: Tanzania, Ugogo, *Stuhlmann* 335 (K, isolecto.!)

Weak-stemmed perennial herb, scrambling over other vegetation or straggling, sometimes in thick masses, 0.3–1.2 m., considerably branched, rooting at the lower nodes; stem and branches striate, sometimes reddish at least below, subglabrous to rather densely pilose with fine appressed hairs. Leaves elliptic to lanceolate, (2–)4–10 × (1–)1.3–4.2 cm., thinly to moderately appressed pilose, paler on the lower surface, acuminate to rather blunt at the apex, cuneate at the base with a distinctly demarcated ± 4–12 mm. petiole. Inflorescence silvery-or yellowish-green to pale mauve, elongating in fruit, 2–26 × 1.2–1.4 cm., subsessile or on a short peduncle to ± 1 cm. long; axis lanuginose. Bracts ovate, 2.5–4 mm., membranous-whitish, ± densely lanuginose, gradually narrowed above to the short arista formed by the excurrent nerve. Bracteoles broadly cordate-ovate, 3–4 mm., lanuginose, more abruptly narrowed to the longer, sharp, sometimes recurved arista. Flowers truncate and finally indurate at the base, attached to the inflorescence-axis by a broadly conical process. Tepals narrowly oblong-lanceolate; 2 outer 4.5–7 mm., densely lanuginose except at the pale-stramineous or pinkish aristate tips, with a rather broad but incurved hyaline margin, 3–5-nerved with the midrib and 2 principal lateral veins frequently branched above; 2 inner slightly shorter and narrower, the hyaline margins as wide as the green 3-nerved central portion, ± pilose or lanate chiefly on the upper part of the central dorsal surface, not aristate; central tepal intermediate, usually lanate at least along one margin as well as above. Perianth and bracteoles falling together in fruit. Stamens 3–4 mm., (? always) red; pseudostaminodes ± 1.5 mm. long, with a short incurved denticulate lobule at the tip and a densely long-fimbriate dorsal scale. Ovary turbinate, often reddish, ±1.5 mm.; style slender, 2–3.25 mm., (? always) red. Ripe capsule and seed not seen.

KENYA. Nairobi District: Nairobi National Park, Mar. 1965, *Kokwaro* 56!; Machakos District: Stony Athi, Aug. 1940, *Nat. Hist. Soc.* 259!; Masai District: Mara Plains, Keekorok [Egalok], Oct. 1958, *Verdcourt & Fraser Darling* 2289!
TANZANIA. Musoma District: Orangi R., Retima [Letema] Pool, June 1962, *Greenway & Turner* 10716!; Mbulu District: ± 1.5 km. N. of Kwa Kuchinja, July 1956, *Milne-Redhead & Taylor* 11191!; Iringa District: Ruaha R. ± 4 km. W. of Mtera bridge, Aug. 1970, *Thulin & Mhoro* 729!
DISTR. **K**4, 6; **T**1, 2, 5, 7; Somalia
HAB Usually in moderately dry country on sandy ground, more rarely on dried-out black cotton soil, in seasonal river beds, river and lake shores, riverine woodland and scrub, also in grazed and valley grassland; 320–1820 m.

SYN. *Achyranthes lanuginosa* Schinz in E.J. 21: 186 (1895); & in P.O.A.C: 174 (1895); F.D.O.-A. 2: 239 (1932); U.K.W.F.: 137 (1974)
Centrostachys schinzii Standley in Journ. Wash. Acad. Sci. 5: 76 (1915)
Achyranthes schinzii (Standley) Cufod. in E.P.A.: 73 (1953)

NOTE. The name *Achyranthes lanuginosa* "Nutt. (1820)" stated by Standley, l.c., to antedate *A. lanuginosa* Schinz, does not appear in the Index Kewensis and I have not been able to trace any description–or any work published by Nuttall in 1820 in which it might have appeared. Standley merely quotes the date.

22. GUILLEMINEA

Kunth in H.B.K., Nov. Gen. Sp. 6: 40, t. 518 (1823), emend. Mears in Sida 3: 137-8 (1967)

Gossypianthus Hook., Ic. Pl. 3, t. 251 (1840)
Brayulinea Small, Fl. S.-East. U.S.: 394 (1903)

Perennial herbs, rarely somewhat woody below, with entire, opposite leaves. Inflorescences sessile, axillary, densely spiciform, often densely fasciculate at the nodes. Flowers ⚥, solitary within the axil of each bract, bibracteolate. Perianth-segments elliptic or ovate, almost entirely delicate and hyaline or firmer with 3 green nerves, free or fused to about halfway; perianth and bracteoles falling with the fruit; bracts persistent. Stamens 5, the filaments fused into a tube, the tube free (not in Africa) or adnate to the perianth-tube; anthers unilocular. Ovary with a single pendulous ovule; style usually short (longer in one West Indian species); stigma shortly bilobed. Capsule thin-walled, bursting irregularly. Seed compressed, firm.

5 species, all natives of the Americas from southern U.S.A. to Argentina; one an increasingly widespread tropical weed.
The genus *Brayulinea* was intended as a replacement for *Guilleminia* Kunth (1823), *non* Necker (1790). Since Necker's names are not recognised as generic under Article 20 of the International Code of Botanical Nomenclature, this substitution is superfluous.

G. densa *(Roem. & Schult.) Moq.* in DC., Prodr. 13 (2): 338 (1852); Mears in Sida 3: 140-144 (1967). Type: "Habitat in America meridionali", Humboldt, *Herb Willdenow* 05066 (B, holo., IDC microfiche 347. 26!)

Prostrate or sometimes decumbent, mat-forming perennial herb with a rootstock considerably thickened for up to ± 5 cm. below the ground and then abruptly more slender, mat from ± 7–70 cm. across. Stems numerous from the base, much-branched, branches opposite (or alternate by reduction of one of the pair), ± densely white-lanate. Leaves variable in size and shape, the lamina narrowly elliptic to broadly ovate, mostly 5–22 × 1.5–14 mm., acute to subacute at the apex, rapidly narrowed below to a broad petiole up to ± 8 mm. long, glabrous or subglabrous on the upper surface, ± densely lanuginose with long, matted, white hairs on the lower surface, especially when young.

FIG. 28. *GUILLEMINEA DENSA*—**1**, flowering branch, × 1.5; **2**, leaf on stem, ×8; **3**, flower in profile, × 16; **4**, flower in plan view, × 16; **5**, part of perianth and androecium flattened out, × 16; **6**. part of perianth and androecium in oblique view, × 16; **7**, gynoecium, × 16. All from *Drummond* 5135. Drawn by Mary Millar Watt.

Inflorescences dense, ovoid, of up to ± 10 flowers, whitish, to ± 6 mm. long, the axis long-pilose; lower flowers frequently minutely pedicellate below the bract, the upper sessile. Bracts hyaline, very delicate and concave, ± 1.5–2 mm., frequently splitting with age, glabrous, persistent; bracteoles similar but slightly shorter. Tepals united for about half their length, densely sinuose-lanuginose, the lobes very delicate, hyaline except for a pale, firmer midrib, whole perianth ± 2–2.5 mm. long at maturity. Staminal tube completely adnate to the perianth-tube, the filaments indicated only by short, triangular teeth; anthers very small, ± 0.25 mm., ovoid. Ovary ellipsoid, firm only at the extreme tip; style very short, ± 0.25 mm. Capsule ellipsoid, ± 1–1.25 mm. Seed compressed-ellipsoid, ± 1 mm. chestnut-brown, faintly reticulate. Fig. 28.

KENYA. Nairobi, Government Road near Jevanjee Gardens, Oct. 1971, *Mwangangi* 1853; Masai District: Namanga Hotel, July 1968, *Agnew* 10164

DISTR. **K** 4, 6; a native of the warmer regions of the Americas from the southern U.S.A. to northern Argentina; introduced into Australia (Queensland) and spreading in S. and tropical Africa, having recently reached Kenya

HAB. Trodden ground; 1500–1600 m.

SYN. *Illecebrum densum* Roem. & Schult., Syst. Veg. ed. 15, 5: 517 (1819)
Guilleminia illecebroides Kunth in H.B.K., Nov. Gen. Sp. 6: 40, pl. 518 (1823). Type: Ecuador, Quito, *Humboldt & Bonpland* (P, holo.!)
Brayulinea densa (Roem. & Schult.) Small, Fl. S.-E.U.S.: 394 (1903); Cavaco in Mém. Mus. Nat. Hist. Nat. Paris, sér. B, 13: 156 (1962); Merxmüller in Prodr. Fl. S.W.A. 33: 9 (1966); Jacot Guillarmod, Fl. Lesotho: 167 (1971)

NOTE. The only amaranth in the Flora region in which the tepals are not free, this species has the mat-forming habit of an *Alternanthera* combined with the woolly inflorescences of an *Aerva*. The wavy, *Gomphrena*-like hairs of the perianth are, however, quite unlike those of *Aerva*.

23. ALTERNANTHERA

Forssk., Fl. Aegypt.-Arab.: 28 (1775)

Annual or perennial herbs, prostrate or erect to floating or scrambling, with entire opposite leaves. Inflorescences of sessile or pedunculate heads or short spikes, axillary, solitary or clustered, bracteate. Flowers ⚥, solitary in the axil of a bract, bibracteolate, bracts persistent, the perianth falling with the fruit, bracteoles persistent or not. Perianth-segments 5, free, equal or unequal, glabrous or furnished with smooth or denticulate hairs. Stamens 2–5, some occasionally without anthers; filaments distinctly monadelphous at the base into a cup or tube, alternating with large and dentate or laciniate to very small pseudostaminodes (rarely these obsolete); anthers unilocular. Ovary with a single pendulous ovule; style short; stigma capitate. Fruit an indehiscent capsule, thin-walled or sometimes corky. Seeds ± lenticular.

A large genus of ± 200 species, chiefly in the New World tropics.

Tepals very dissimilar in form:
  Abaxial tepals (Fig. 29/7, 8) very long-aristate (the awn ± ⅓ of the total tepal length), with tufts of barbellate hairs near the basal angles; adaxial tepal (Fig. 29/9) strongly denticulate .... 1. *A. pungens*
  Abaxial tepals (Fig. 29/2) only shortly aristate (the awn less than ¼ of the total length), with barbellate hairs in the basal half or more; adaxial tepal (Fig. 29/3) almost entire .... 2. *A. caracasana*
Tepals all similar in form, subequal or the inner 2 slightly shorter:
  Outer tepals prominently 3-nerved in the lower half, with numerous barbellate hairs .... 3. *A. tenella* var *bettzickiana*
  Outer tepals 1-nerved with the prominent midrib only:

Fruit yellowish with tumid margins, commonly only about half the length of the usually 3–4 mm. perianth-segments .... 4. *A. nodiflora*
Fruit, dark, thin-margined with only a narrow yellowish rim, almost as long or sometimes exceeding the 1.5–2.5 mm. perianth-segments .... 5. *A. sessilis*

1. **A. pungens** *Kunth* in H.B.K., Nov. Gen. Sp. 2: 206 (1818); U.K.W.F.: 137 (1974). Type: Colombia, R. Orinoco, Maypur waterfall, *Humboldt & Bonpland* (P, iso.!)

Prostrate, mat-forming perennial herb with a stout vertical rootstock, also rooting at the lower nodes, much branched from the base outwards, mats up to ± 1 m. across. Stem and branches terete, striate, stout to more slender, ± densely villous with long white hairs but frequently glabrescent with age. Leaves broadly rhomboid-ovate to broadly elliptic or obovate, 1.5–4.5 × 0.3–2.7 cm., rounded to subacute at the apex with a mucro which in the young leaves is often fine and bristle-like, narrowed below to a petiole up to 1 cm. long, glabrous or thinly appressed pilose on both surfaces, especially on the lower surface of the primary venation. Inflorescences sessile, axillary, solitary or more commonly 2–3 together, globose to shortly cylindrical, 0.5–1.5 cm. long and 0.5–1 cm. wide; bracts membranous, white or stramineous, deltoid-ovate, 4–5 mm., glabrous, marginally ciliate or dorsally pilose, distinctly aristate with the excurrent midrib, ± denticulate around the upper margin; bracteoles similar but smaller, 3–4 mm., falling with the fruit. Tepals extremely dissimilar; 2 outer (abaxial) deltoid-lanceolate, 5 mm., denticulate above, very rigid, 5-nerved below (the intermediate pair much shorter and finer), outer 2 nerves meeting above to join the pungently excurrent midrib, which forms a long arista ± ⅓ the length of the entire tepal; inner (adaxial) tepal oblong, flat, 3 mm., blunt and strongly dentate at the apex, 3-nerved below but the nerves meeting well below the apex and the apical mucro short and fine; lateral tepals ± 2mm., sinuate in side view with the two sides of the lamina connivent and denticulate above, sharply mucronate; abaxial and adaxial tepals with small tufts of glochidiate and barbed whitish bristles about the basal angles, the lateral tepals each with a large tuft about the centre of the midrib. Stamens 5, all with anthers, at anthesis subequalling or slightly exceeding the ovary and style, the alternating pseudostaminodes broad, subquadrate or shorter, entire to dentate. Ovary compressed, squat; style very short, as wide as or wider than long. Fruit roundish, rounded to retuse above. ± 2 mm. Seed discoid, ± 1.25 mm., brown, shining, faintly reticulate. Fig. 29/7–10.

UGANDA. W. Nile District: Nebbi, Sept. 1940, *Purseglove* 1058!; Teso District: Serere, Dec. 1931, *Chandler* 251!; Mengo District: Kampala, Oct. 1929, *Liebenberg* 1067!
KENYA. Northern Frontier Province: Isiolo–Marsabit road, bridge over Serolevi R., Mar. 1963, *Bally* 12493!; Nairobi, by railway, Oct. 1929, *McDonald* in *A.D.* 1345!; Kilifi District: Malindi, May 1960, *Rawlins* 839!
TANZANIA. Musoma District: Serengeti, Seronera Lodge grounds, Apr. 1961, *Greenway* 10099!; Lushoto District: Korogwe, Nov. 1962, *Archbold* 7!; Kilosa District: Msange Mbuga, June 1973, *Greenway & Kanuri* 15266!; Zanzibar, Jan. 1962, *Faulkner* 2997!
DISTR. **U**1–4; **K**1–5, 7; **T**1–3, 5–7; **Z**; a native of tropical America now widespread as a weed in the tropics and subtropics of both Old and New Worlds
HAB. Appears to thrive best in bare, frequently heavily trodden places along tracks and roadsides, also on river banks, in *Cynodon* grassland, and on eroded ground in forests, on a variety of soils from sandy alluvium to heavy clay and black cotton soil; 0–1520 m.

SYN. *Achyranthes repens* L., Sp. Pl.: 205 (1753). Type: Eltham Gardens, ex Herb. Dillenius (OXF, lecto.)*

*Fide Melville in K.B., 13: 172–3 (1958). The specimens under the name *Achyranthes repens* in Linnaeus' herbarium at Stockholm appear from IDC microfiche 101.9, 11 to be *A. repens* Gmel. -i.e. *A. sessilis.*

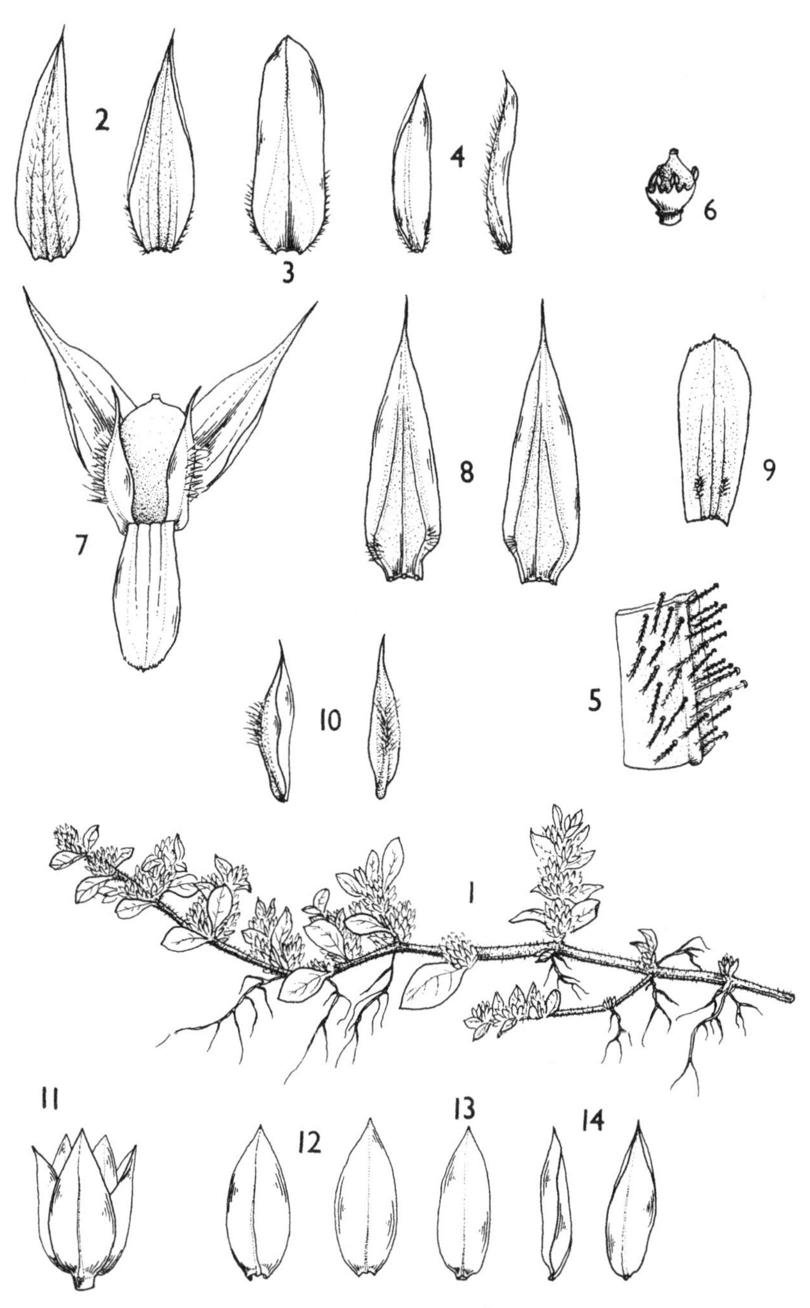

FIG. 29. *ALTERNANTHERA CARACASANA*—**1**, flowering branch, × $\frac{2}{3}$; **2**, outer tepals, × 8; **3**, median tepal, × 8; **4**, inner tepals, × 8; **5**, hairs from inner tepals, × 24; **6**, androecium and gynoecium, × 8. *A. PUNGENS*—**7**, flower opened up, × 8; **8**, outer tepals, × 8; **9**, median tepal, × 8; **10**, inner tepals, × 8. *A. SESSILIS*—**11**, flower, × 8; **12**, outer tepals, × 8; **13**, median tepal, × 8; **14**, inner tepals, × 8. 1–6, from *Hooper & Townsend* 1649; 7–10, from *Archbold* 7; 11–14, from *Bally & Smith* 14836. Drawn by Christine Grey-Wilson.

Illecebrum achyrantha L., Sp. Pl., ed. 2: 299 (1762). Type: as for above
Alternanthera achyrantha (L.) Sweet, Hort. Suburb. London: 48 (1818)
A. echinata Sm. in A. Rees, Cyclop. 39, Add. & Corrig.: sub. Alternanthera No. 10 (1819); F.T.A. 6(1): 74 (1909); F.D.O.-A. 2: 244 (1932). Type: Uruguay, Montevideo, *Commerson* s.n. (Smith Herbarium No. 426.21, LINN, holo.!, IDC microfiche 215. 19, not indexed under this name)
A. repens (L.) Link, Enum. Pl. Hort. Berol. alt. 1: 154 (1821), *non* Gmel., Syst. Nat., ed. 13, 2(8): 106 (1791); F.P.S. 1: 116 (1950); Hauman in F.C.B. 2: 76 (1951); E.P.A.: 74 (1953); Cavaco in Mém. Mus. Nat. Hist. Nat. Paris, sér. B, 13: 161 (1962)

2. **A. caracasana** *Kunth* in H.B.K., Nov. Gen. Sp. 2: 205 (1818). Type: Venezuela, near Caracas and Chacao, *Humboldt & Bonpland* in *Herb. Willdenow* 5029 (B, holo., IDC microfiche 345. 5 !)

Prostrate, mat-forming perennial herb with a stout vertical rootstock, also rooting at the lower nodes, much branched from the base outwards, mats up to ± 1 m. across. Stem and branches terete, striate, stout to slender, ± densely villous with long white hairs. Leaves broadly ovate to broadly elliptic or obovate, 0.8–3.2 × 0.4–1.5 cm., rounded to subacute at the apex, sharply mucronate, narrowed below to a distinct petiole up to ± 1 cm. long, glabrous to thinly long-pilose on both surfaces, especially about the base. Inflorescences sessile, axillary, solitary or 2–3 together, usually shortly cylindrical, 0.5–1.5 cm. long and 0.4–0.7 cm. wide; bracts membranous, white or stramineous, deltoid-ovate, 3–3.5 mm., glabrous or slightly pilose at the basal margins, distinctly aristate with the excurrent midrib, entire; bracteoles lanceolate, sometimes falcate, commonly somewhat wider on one side of the midrib than the other, 3–3.5 mm., glabrous or slightly pilose along the keel, subentire. Tepals extremely dissimilar; outer 2 (abaxial) deltoid-lanceolate, 4–4.5 mm., very rigid, 3-nerved below with the laterals meeting the midrib ± $\frac{2}{3}$ of the way up, midrib excurrent to form an arista less than $\frac{1}{4}$ the length of the tepal; inner (adaxial) tepal oblong, flat, 3–4 mm., entire or faintly denticulate at the apex, 3-nerved with the laterals meeting the midrib ± $\frac{2}{3}$ up, finely mucronate; lateral tepals 2.75–3 mm., sinuate in side view with the two sides connivent above, entire or almost so, sharply mucronate; abaxial and adaxial tepals with glochidiate and barbed whitish hairs to about halfway or more, the laterals with a long tuft in ± the centre half of the midrib. Stamens 5, all with anthers, at anthesis subequalling or slightly exceeding the ovary and style, the alternating pseudostaminodes narrowly triangular-subulate, subequalling or slightly shorter than the filaments. Ovary compressed, squat; style very short, about as long as wide. Fruit roundish, rounded to retuse above, ±2 mm. Seed discoid, ±1.25 mm., brown, shining, faintly reticulate. Fig. 29/1–6.

UGANDA. Kigezi District: Kabale, June 1949, *Purseglove* 2897!
KENYA. Laikipia District: 43 km. from Nyahururu [Thomsons Falls] on road to Nanyuki, Apr. 1977, *Hooper & Townsend* 1649!; Kericho District: Sambret-Timbilil, Sept. 1961, *Kerfoot* 2820!; Masai District: Ol Choro Orogwe Ranch, July 1961, *Glover, Gwynne, Samuel & Tucker* 2020!
TANZANIA. Mpanda railway station, July 1968, *Sanane* 229!; Morogoro District: Mkata Ranch, Mar. 1974, *Wingfield* 2629!; Chunya District: Kipembawe [Kepembawe] Game Camp, Mar. 1965, *Richards* 19824!
DISTR. U2; **K**1, 3–6; **T**2, 4, 6, 7; a native of tropical America, introduced as a weed in Zambia, Zimbabwe, Botswana, Angola and South Africa
HAB. By roadsides, trodden ground along paths, waste ground, in derelict cultivation and similar places, mostly on sandy soil; 430–1870 m.

SYN. *Illecebrum peploides* Schultes, Syst. Veg., ed. 15, 5: 517 (1819). Type: Santo Domingo, *Humboldt & Bonpland* in *Herb. Willdenow* 5064, (B, holo., IDC microfiche 347. 21!)
Telanthera caracasana (Kunth) Moq. in DC., Prodr. 13(2): 370 (1849)
Alternanthera peploides (Schultes) Britton in Britton & Brown, Bot. Puerto Rico & Virgin Is. 2: 279 (1924); U.K.W.F.: 137 (1974)

3. **A. tenella** *Colla* in Mem. R. Accad. Sci. Torino 33: 131, t. 9 (1828)

var. **bettzickiana** *(Regel) Veldk.* in Taxon 27: 313 (1978). Type: cultivated material from St. Petersburg Botanic Garden (LE, holo.!)

Erect or ascending bushy perennial herb (commonly cultivated as an annual), ± 5–45 cm. tall; stem and branches villous when young but soon glabrescent, older parts terete, younger bluntly quadrangular. Leaves narrowly or more broadly elliptical to oblanceolate or rhombic-ovate, acute to acuminate at the apex, attenuate into a slender indistinctly demarcated petiole below, thinly furnished with fine whitish hairs to subglabrous, often reddish or purple suffused and not rarely variegated. Heads axillary, sessile, usually solitary, globose to ovoid, 4–6 mm. in diameter; bracts pale, deltoid-ovate, ± 2 mm., glabrous, lacerate-margined, aristate with the excurrent midrib; bracteoles similar but slightly shorter. Tepals white, lanceolate to oblong-elliptic, 3.5–4 mm., acute, mucronate with the excurrent midrib; outer 2 prominently 3-nerved below and darker in the nerved area, with a line of whitish minutely barbellate hairs on each side of this area, the hairs becoming denser towards the base of the tepal; inner 2 slightly shorter, narrower and less rigid, mostly 1–2-nerved; central tepal intermediate. Stamens 5, at anthesis much exceeding the ovary and style, the alternating pseudostaminodes subequalling the filaments plus anthers, narrowly oblong, laciniate at the tip. Ovary strongly compressed, obpyriform, 0.6 mm. long; style about the same length. Ripe fruit and seeds not seen.

UGANDA. Masaka District: near Kisasa, 2.5 km. W. of Bunado, May 1972, *Lye* 6950!; Masaka, May 1972, *Katende* 1666!

TANZANIA. Shinyanga District: Mwadui, Oct. 1971, *Batty* 1461!; Lushoto District: Amani (cult.), Mar. 1906, *Braun* in *Herb. Amani* 1116!; Uzaramo District: Msasani–Dar es Salaam. State Lodge No. 2, Sept. 1972, *Ruffo* 479!

DISTR. U4; T1, 3, 6; said to be a native of S. America, probably Brazil, but no certainly wild material has been seen and this is probably conjecture

HAB. In and near cultivation, by roadsides–probably a short-term "escape"

SYN. *Telanthera bettzickiana* Regel in Ind. Sem. Hort. Petrop. 1862: 28 (1862) & in Gartenflora 11: 178 (1862)
Alternanthera bettzickiana Nichols., Ill. Dict. Gard. 1: 59 (1884), *nom. subnud.*
A. bettzickiana (Regel) Voss in Vilm., Blumengart., ed. 3, 1: 689 (1895), sphalm. "bettzichiana"; Schinz in Pflanzenfam., ed. 2, 16C: 75 (1934); Hauman in F.C.B. 2: 79 (1951)
[*A. amoena* sensu Peter in F.D.O.–A. 2: 244 (1932),? an (Regel) Voss]
A. ficoidea (L.) Griseb. var. *bettzickiana* (Nichols.) Backer in Fl. Males., Ser. 1, 4: 93 (1949)

NOTE. The plant is widespread as a decorative border plant because of its ± variegated foliage, and is believed to be a cultigen of *A. tenella* Colla (*A. ficoidea* auctt.). As will be seen above, ripe seeds (normally abundant in the genus) have not been seen, and var. *bettzickiana* is propagated readily by cuttings.

4. **A. nodiflora** *R. Br.*, Prodr. Fl. Nov. Holl.: 417 (1810); F.T.A. 6(1): 73 (1909); F.D.O.–A. 2: 244 (1932); F.P.S. 1: 115 (1950); Hauman in F.C.B. 2: 74 (1951); E.P.A.: 74 (1953); Cavaco in Mém Mus. Nat. Hist. Nat. Paris, sér. B, 13: 160 (1962). Type: Australia, no details, *Robert Brown* 3066 (K, lecto.!)

(Annual? or) perennial herb with a stout vertical rootstock, prostrate to ascending or erect, frequently rooting at the lower nodes, slightly to considerably branched from the base outwards, mostly 12–40 cm. Stem and branches green to purplish brown, terete below and ± tetragonous above, with a narrow line of whitish hairs down each side of the stem and branches (at least when young) and tufts of white hairs in the branch and leaf-axils, otherwise glabrous. Leaves linear to narrowly elliptic, 2–7 × 0.3–1.2 cm., acute to subacute at the apex, attenuate below, glabrous or occasionally thinly pilose when young; petiole indistinct or absent. Inflorescences sessile, axillary, solitary or in clusters of up to ± 5, globose, ± 7–10 mm. in diameter, often meshing with one another and those of the opposite leaf-axil to form a dense mass; bracts scarious, white, ovate-acuminate, (2–)2.25 × 2.75 mm., mucronate with the excurrent pale midrib, glabrous; bracteoles ovate-lanceolate, slightly longer mucronate, (2–)2.25–2.5 mm., glabrous. Tepals ovate-acuminate, tapering from below the middle (in smaller forms less long-pointed and tapering from about the middle), (2.75–)3–4 mm., white to pink tinged, glabrous, shortly mucronate with the excurrent nerve, the margins often obscurely lacerate-denticulate. Stamens 5 (2 filaments anantherous), at anthesis subequalling the ovary and style, the alternating pseudostaminodes resembling the filaments but a little shorter. Ovary strongly compressed, roundish; style very short. Fruit obcordate, 2–2.5 mm. wide, 1.75–2 mm. long, about half as long as the perianth, strongly compressed, glabrous, with broad tumid yellow margins and darker over the seed. Seed discoid, ± 1.25–1.5 mm., brown, shining, faintly reticulate.

TANZANIA. Tabora District: Malongwe, Kisengi, Aug. 1952, *Bally* 8303!; Dodoma District: Imagi Hill, Jan. 1962, *Polhill & Paulo* 1284!; Iringa District: Mbagi, by Mwagusi R., Jan. 1966, *Richards* 21030!

DISTR. T3–5, 7; throughout Australia and tropical and South Africa; naturalised in Taiwan, Japan and probably elsewhere

HAB. Always in damp (at least seasonally) ground, in gravel pit, by pools and rivers, one record from irrigated land; mostly on sand, but also on cracking clay; 760-1370 m.

SYN. *A. sessilis* (L.) DC. var. *nodiflora* (R. Br.) Kuntze, Rev. Gen. 2: 540 (1891)

5. **A. sessilis** *(L.) DC.*, Cat. Hort. Monsp.: 77 (1813); P.O.A.C: 174 (1895); F.P.N.A. 1: 138 (1948); F.P.S. 1: 115 (1950); Hauman in F.C.B. 2: 73 (1951); E.P.A.: 75 (1953); Cavaco in Mém. Mus. Nat. Hist. Nat. Paris, sér. B, 13: 158 (1962); U.K.W.F.: 137 (1974). Type: *Hermann Herbarium* Vol. 2, p. 78 (BM, lecto.!)

Annual or usually perennial herb; in drier situations with slender more solid stems, prostrate, decumbent or erect, ± much branched, to ± 30 cm.; in wetter places ascending or most commonly prostrate with stems ± 0.1–1 m. long, rooting at the nodes, ± fistular, with numerous lateral branches; when floating very fistular, the stems attaining a metre or more in length and over 1 cm. thick, with long clusters of whitish rootlets at the nodes. Stem and branches green to pink or purplish, with a narrow line of whitish hairs down each side of the stem and branches (at least when young) and tufts of white hairs in the branch and leaf axils, otherwise glabrous, striate, terete below, ± tetragonous above. Leaves extremely variable in shape and size, linear-lanceolate to oblong, oval or obovate-spathulate, 1–9(–15) × 0.2–2 (–3) cm., blunt to shortly acuminate at the apex, cuneate to attenuate at the base, glabrous or thinly pilose, especially on the lower surface of the midrib; petiole obsolete to ± 5 mm. Inflorescences sessile, axillary, solitary or in clusters of up to ± 5, subglobose (slightly elongate in fruit), ± 5 mm. in diameter; bracts scarious, white, deltoid-ovate, ± 0.75–1 mm., mucronate with the excurrent pale midrib, glabrous; bracteoles similar, 1–1.5 mm., also persistent. Tepals oval-elliptic to lanceolate-ovate, equal, 1.5–2.5 mm., acuminate to rather blunt, white to pink-tinged, glabrous, shortly but distinctly mucronate with the stout excurrent midrib, the margins often obscurely lacerate-denticulate. Stamens 5 (2 filaments anantherous), at anthesis subequalling the ovary and style, the alternating pseudostaminodes resembling the filaments but usually somewhat shorter. Ovary strongly compressed, roundish, style extremely short. Fruit glabrous, obcordate or cordate-orbicular, 2–2.5 mm. long, subequalling or slightly shorter than the perianth, strongly compressed, dark brownish with a narrow, pale yellowish, somewhat thickened margin. Seed discoid, ± 0.75–1 mm., brown, shining, faintly reticulate. Fig. 29/11–14.

UGANDA. Karamoja District: Kaleri [Napore] Hills, Karenga, Dec. 1972, *J. Wilson* 2176!; Busoga District: W. of Busia, Aug. 1964, *Tweedie* 2887!; Mengo District: Port Kibanga, Aug. 1914, *Dummer* 1024!

KENYA. Northern Frontier Province: Marsabit, Lake Paradise, Mar. 1959, *Bogdan* 4787!; Nairobi, Feb. 1915, *Dummer* 1953!; Teita District: Galana R., W. of Lugard Falls, Jan. 1967, *Greenway & Kanuri* 13041!

TANZANIA. Mwanza District: Bwiru, Jan. 1954, *Tanner* 1526!; Lushoto District: Korogwe, Oct. 1963, *Archbold* 306!; Iringa District: Great Ruaha R., Msembe, Dec. 1962, *Richards* 17365!; Zanzibar I., Zingwe-Zingwe, Jan. 1929, *Greenway* 1092!; Pemba I., Chake Chake, Aug. 1929, *Vaughan* 559!

DISTR. **U**1–4; **K**1–4, 6, 7; **T**1–8; **Z**; **P**; very widespread in the tropics and subtropics of both Old and New Worlds

HAB. Most commonly by or in water, also in damp forests and abandoned cultivation, on many kinds of soil from poor sandy to loam or black cotton; 0–1840 m.

SYN. *Gomphrena sessilis* L., Sp. Pl.: 225 (1753)
Illecebrum sessile (L.) L., Sp. Pl., ed. 2: 300 (1762)
Alternanthera "achyranth." Forssk., Fl. Aegypt-Arab.: lix (1775); F.T.A. 6(1): 73 (1909): F.D.O.-A. 2: 243 (1932), subnom. *"achyranthoides"*
Alternanthera repens Gmel., Syst. Nat., ed. 13, 2(1): 106 (1791). Type: *Forsskål* Cent. 1, 100 (C)

NOTE. Said on two herbarium sheets (*Archbold* 306 and *Njogu* in *E.A.H.* 13834) to be used as a vegetable. The author has seen this plant tied in bundles in Sri Lanka for salad use, as watercress is in Britain.

24. GOMPHRENA

L., Sp. Pl.: 224 (1753) & Gen. Pl., ed. 5: 105 (1754)

Annual or occasionally perennial herbs with entire opposite leaves. Inflorescences terminal or axillary, capitate or spicate, solitary or glomerate, often subtended by a pair of sessile leaves, bracteate with the bracts persistent in fruit, the axis frequently thickened. Flowers ⚥, each solitary in the axil of a bract, bibracteolate; bracteoles (in African species) laterally compressed, carinate, often ± winged or cristate along the dorsal surface of the midrib, deciduous with the fruit. Tepals 5, erect, free or almost so, ± lanate dorsally, at least the inner 2 usually ± indurate at the base in fruit. Stamens 5, monadelphous, the tube shortly 5-dentate with entire to very deeply bilobed teeth; anthers unilocular. Ovary with a single pendulous ovule; style short or long; stigmas 2, suberect or ± divergent, filiform to very short. Fruit a thin-walled irregularly rupturing capsule. Seed ovoid, compressed.

A rather large genus of about 90 species, centred on tropical America but also with several Australian representatives.

Bracteoles with a small dorsal crest confined to about the upper one-third of the dorsal surface of the midrib .... 1. *G. celosioides*
Bracteoles with a large, conspicuous dorsal crest extending from the apex almost to the extreme base .... 2. *G. globosa*

1. **G. celosioides** *Mart.* in Nov. Act. Acad. Caes.-Leop. Carol., Nat. Curios, 13: 301 (1826); Sandwith in K.B. 1: 29 (1946); Milne-Redh. in K.B. 2: 23 (1947); Hauman in F.C.B. 2: 79 (1951); Cavaco, in Mém. Mus. Nat. Hist. Nat. Paris, sér. B, 13: 165 (1962); F.P.U.: 101 (1962); U.K.W.F.: 139 (1974). Type: Brazil, *Sellow* (K, iso.!)

Perennial herb, prostrate and mat-forming to ascending or erect, ± 7–30 cm., much branched from the base and also above; stem and branches striate, often sulcate, green to reddish, when young usually ± densely furnished with long white lanate hairs, ± glabrescent with age. Leaves narrowly oblong to oblong-elliptic or oblanceolate, ± 1.5–4.5 (–8) × 0.5–1.3(–1.8) cm., obtuse to subacute at the apex, mucronate, narrowed to a poorly demarcated petiole below, the pair of leaves subtending the terminal inflorescence more abruptly narrowed and sessile, oblong or lanceolate-oblong; all leaves glabrous or thinly pilose above, thinly to densely furnished with long whitish hairs on the margins and lower surface, sometimes ± lanate on both surfaces. Inflorescences sessile above the uppermost pair of leaves, white, at first subglobose and ± 1.25 cm. in diameter, finally elongate and cylindrical, ± 4–7 cm. long with the lower flowers falling; axis ± lanate; bracts deltoid-ovate, 2.5–4 mm., shortly mucronate with

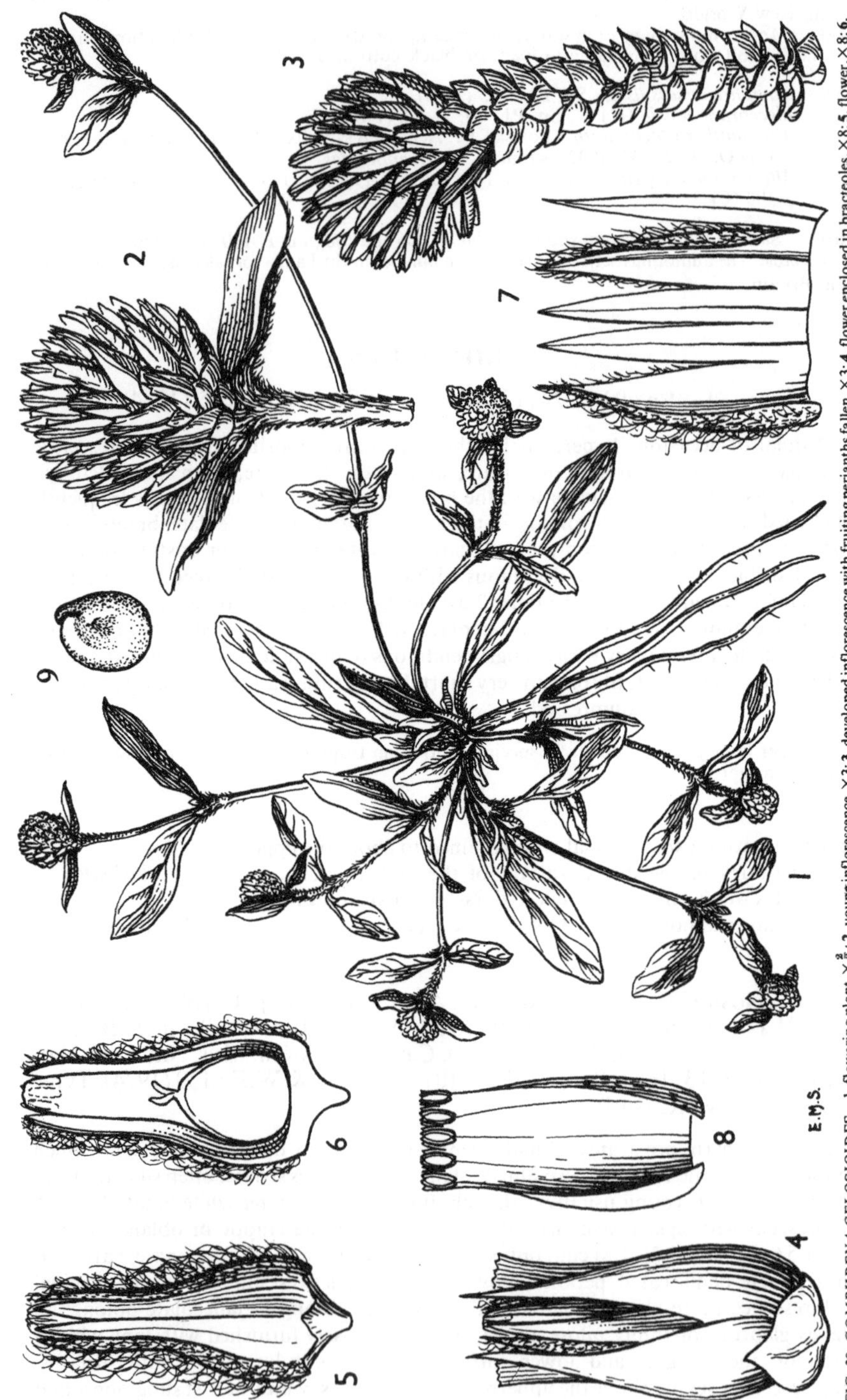

FIG. 30. *GOMPHRENA CELOSIOIDES*.—1. flowering plant, $\times\frac{2}{3}$; **2.** young inflorescence, ×3; **3.** developed inflorescence with fruiting perianths fallen, ×3; **4.** flower enclosed in bracteoles, ×8; **5.** flower, ×8; **6.** longitudinal section of flower, ×8; **7.** perianth opened out, × 8; **8.** staminal tube opened out, × 8; **9.** seed, × 8. 1, 3, from *Hancock* 41; 2, 4, 9, from *Verdcourt* 351. Drawn by Margaret Stones.

the excurrent midrib; bracteoles strongly laterally compressed, navicular, ± 5–6 mm., mucronate with the excurrent midrib, furnished along about the upper one-third of the dorsal surface of the midrib, with an irregularly dentate or subentire wing. Tepals ± 4.5–5 mm., narrowly lanceolate, 1-nerved, the outer 3 ± flat, lanate only at the base, nerve thick and greenish below, thinner and excurrent in a short mucro at the tip; inner 2 sigmoid in lateral view at the strongly indurate base, densely lanate almost to the tip, slightly longer. Staminal tube subequalling the perianth, the 5 teeth deeply bilobed with obtuse lobes subequalling the ± 0.75 mm. anthers which are set between them; pseudostaminodes absent. Style and stigmas together ± 1 mm.; style very short; stigmas divergent. Capsule shortly compressed-pyriform, ± 1.75 mm. Seed compressed-ovoid, ± 1.5 mm., brown, faintly reticulate, shining. Fig. 30.

UGANDA. W. Nile District: Pakwach, Apr. 1940, *Eggeling* 3922!; Teso District: Serere, Nov. 1931, *Chandler* 33!; Mengo District: Kampala, *Liebenberg* 1048!
KENYA. Northern Frontier Province: Wamba, Dec. 1958, *Newbould* 3144!; Nairobi, new racecourse, July 1930, *Napier* 232!; Lamu District: 3 km. S. of Witu towards Kipini, Mar. 1977, *Hooper & Townsend* 1168!
TANZANIA. Mwanza District: Kalemera, Oct. 1952, *Tanner* 1105!; Tanga District: Magunga Estate, June 1953, *Faulkner* 1199!; Songea, Feb. 1976, *Milne-Redhead & Taylor* 8751!; Zanzibar I., Marahubi, July 1963, *Faulkner* 3212!
DISTR. **U** 1, 3, 4; **K** 1–5, 7; **T** 1–8; **Z**; a native of S. America (S. Brazil, Paraguay, Uruguay and Argentina) which has spread with tremendous rapidity during the present century to become a widely distributed weed in the tropics and subtropics generally. The oldest gathering in our region which has been seen is *Zimmermann* in *Herb. Amani* 6522, collected in **T** 3 in Aug. 1916.
HAB. Particularly common on trodden ground by paths and roadsides and in poor inland and coastal grassland, also as a weed of cultivation, on waste ground, by lake margins and in seasonally flooded areas near rivers; on all kinds of soil from poor sandy/gravelly soil to rich brown loam and black cotton; 0–2130 m.

SYN. *G. globosa* L. subsp. *africana* Stuchlik in Beih. Bot. Centralbl. 30(2): 396 (1913). Type: *Leendertz* 1888 (K, syn.!; Stuchlik cited the locality, "P.P. Rust", as the collector)
[*G. decumbens* sensu Stuchlik in Burtt Davy, Fl. Pl. & Ferns Transv. 1: 185, fig. 20 (1926), *non* Jacq.]
G. alba Peter in F.R. Beih. 40(2), Descr.: 24 (1932). Types: Tanzania, S. Pare, Buiko, *Peter* 12420 & Kilimanjaro, New Moshi, *Peter* 42212 (both B, syn.†)*
G. celosioides Mart. forma *villosa* Suesseng. in F.R. 42: 57(1937). Type: Namibia, *Dinter* 8002 (K, iso.!)

NOTE. According to Mears (Taxon 29: 86–7, 1981), this widespread weed has been misidentified as *G. celosioides* Mart., and is really *G. decumbens* Jacq., which in turn he avers should now be called *G. serrata* L. He claims that true *Gomphrena celosioides* is restricted to South America except in cultivation. However, the original description and (very good) plate of *G. decumbens* indicate a plant with (1) an annual root (2) bracteoles with a serrate keel reaching almost to the base and (3) heads sessile in groups of three on the stems and principal branches. None of these characters can possibly apply to the weedy African species. I cannot separate the African plant from isotypes of *G. celosioides* which I have seen, and shall continue to use this name for it. The resurrection of long-lapsed names such as *G. serrata*, based on short diagnoses without the availability of a genuine type, is not to be recommended.

2. **G. globosa** *L.*, Sp. Pl: 224 (1753); P.O.A.C: 174 (1895); F.T.A. 6(1): 75 (1909); F.D.O.-A. 2: 245 (1932); U.O.P.Z.: 277 (1949); Hauman in F.C.B. 2: 80 (1951); E.P.A.: 75 (1953); Cavaco in Mém. Mus. Nat. Hist. Nat. Paris, sér. B. 13: 167 (1962). Type: *Herb. Hort. Cliffort.*, specimen ". . . capitulis argenteis . . ." (BM, lecto.!)

Annual herb, decumbent or erect, branched from the base and also above, ± 15–60 cm.; stem and branches striate or sulcate, ± densely clothed with appressed white hairs at least when young. Leaves broadly lanceolate to oblong or elliptic-oblong, 2.5–12(–15) × 2–4(–6) cm., narrowed to an ill-defined petiole below, thinly pilose on both surfaces, the pair of leaves subtending the terminal inflorescence sessile or almost so, broadly

*Specimen *Peter* 24354, from Useguha, is still extant; it was for some reason not cited by Peter.

ovate to subcordate-ovate. Inflorescences sessile above the uppermost pair of leaves, usually solitary, globose or depressed-globose, rarely ovoid, ±2 cm. in diameter, whitish to pinkish or deep red; bracts deltoid-ovate, 3–5 mm., mucronate with the shortly excurrent midrib; bracteoles strongly laterally compressed, navicular, ±8–12 mm., mucronate with the excurrent midrib, furnished from the apex almost to the base of the dorsal surface of the midrib with a broad irregularly dentate crest. Tepals similar to those of *G. celosioides* but longer (6–6.5 mm.), the outer more lanate and the inner less markedly indurate at the base. Staminal tube subequalling the perianth, the 5 teeth deeply bilobed with obtuse lobes subequalling the anthers; pseudostaminodes absent. Style and stigmas together ± 2.5 mm.; stigmas divergent, subequalling or slightly longer than the style. Capsule oblong-ovoid, compressed, ±2.5 mm. Seeds compressed-ovoid, ±2 mm., brown, almost smooth, shining.

UGANDA. Teso District: roadside through Kamod Swamp, Serere, July 1934, *Synge* 746!; Masaka, cultivated ornamental, May 1972, *Lye* 6838!
KENYA. Kiambu District: Muguga, cult. Hort. E.A.A.F.R.O., Feb. 1961, *Greenway* 9834!; Kilifi District; Kibarani, Feb. 1946, *Jeffery* K468! & Rabai Hills, *W.E. Taylor!*
TANZANIA. Ufipa District: Rukwa, Milepa, Mar. 1949, *Burnett!*; Tunduru, *Allnutt* 1!

NOTE. A native of tropical America, long cultivated in the warmer parts of the world and its exact native area obscured. The status of one Tanzanian specimen cited, *Burnett* s.n., is not given, of the other "common bush flower". This latter sounds dubious, but it is remarkable in view of the success of *G. celosioides* that although the present species may occur as a garden stray or outcast, in the Old World at least it rarely if ever becomes truly naturalised. The Jeffery Kenya specimen, which observes that the plant was common in grass on sand, may be a case of naturalisation. A note on this sheet observes that a chest medicine is prepared from the roots.

25. IRESINE

P. Browne, Civ. & Nat. Hist. Jamaica; 358 (1756) nom. conserv.

Cruicita L., Sp. Pl., ed. 2: 179 (1762)
Xerandra Raf., Fl. Tellur. 3: 43 (1836)
Ireneis Moq. in DC., Prodr. 13 (2): 349 (1849)

Annual or perennial herbs, shrubs or subshrubs, sometimes scandent, rarely small trees, with opposite leaves and branches. Leaves simple, entire or ± incised. Flowers androgynous, polygamous or dioecious, bibracteolate, inflorescence frequently paniculate. Perianth-segments 5, free, frequently pilose or lanate. Stamens 5, shortly monadelphous at the base; pseudostaminodes occasionally well developed, more commonly short or obsolete; anthers unilocular. Ovary with a single pendulous ovule; stigmas 2–3, elongate, or sometimes capitate in functionally ♂ flowers. Capsule membranous, indehiscent. Embryo annular, endosperm present.

About 80 species, native to the New World and principally in the tropics.

I. herbstii *Lindl.*, Gard. Chron. 1864: 654 (1864). Type: cultivated material, England, origin R. Plate (K, holo.!)

Erect or ascending perennial herb, to ± 1.5 m.; stem and branches rather succulent, sulcate when dry, with rather sparse fine multicellular hairs. Leaves variegated, from purplish with pink veins to green with yellow veins, orbicular to ovate-orbicular, shortly cuneate to truncate at the base, mostly deeply retuse at the apex (sometimes merely rounded), on the lower surface sparingly furnished with short frequently golden-glistening hairs. Flowers unisexual on separate plants, sessile, in many-flowered much-branched panicles up to ± 20 cm. long. Perianth-segments oblong, blunt or subacute, 1–1.25 mm., 3-nerved, glabrous (but lanate hairs often occurring at the axis between the tepals and the bracteoles).

KENYA. S. Nyeri District: Ragati Forest Station, in garden, 2 Oct. 1964, *Gillett* 16276!
TANZANIA. Lushoto District: Amani nursery, 5 Apr. 1932, *Greenway* 2959!

NOTE. A native of S. America widely cultivated as a decorative in the tropics (and under glass in cooler climates). In some areas in the Old World it appears to be ± naturalised, and may be looked for as such in E. Africa.

Excluded species

Psilodigera spicata Suesseng was described by Suessenguth in Mitt. Bot. Staats., München 1: 109 (1952), and later transferred by Cavaco to *Psilotrichum* in Mém. Mus. Nat. Hist. Nat. Paris, sér. B., 13: 112 (1962). It has been shown elsewhere that the type specimen, *F.A. Rogers* 3304b, labelled as being collected at Tanga, is *Saltia papposa* (Forssk.) Moq. and was certainly incorrectly localised (see Townsend in K.B. 28: 143 (1973)).

Kyphocarpa petersii *Lopr.* in E.J. 27: 43 (1899) was recorded by Peter in F.D.O.-A. 2:229 (1938) from Tanzania (**T6**, Kilosa District, Kidete) on the basis of two of his own gatherings - viz. *Peter* 32734 & 32800. *K. petersii* was described from Tete, Mozambique from a specimen seen at Berlin by Lopriore, *Peters* s.n. The holotype is no longer to be found at Berlin, presumably having been destroyed in World War II; *Peter* 32734 & 32800. are similarly missing, and it is not possible to be certain of their identity. However, to judge from Lopriore's original description it seems to me that Peter's specimens were probably misidentified *Centemopsis gracilenta*. He does not otherwise mention this species, which is recorded from **T4**, 5 & 7 and is certainly to be expected in **T6**.

INDEX TO AMARANTHACEAE

	A	B	C
Dipsacaceae (£1.00)	92	183	132
Dipterocarpaceae	26	125	69
Droseraceae (£0.50)	62	81	25
Ebenaceae	101	159	122
Elatinaceae (£0.50)	22	126	28
Ericaceae	96	154	121
Eriocaulaceae	186	16	163
Erythroxylaceae (£1.40)	32	94	82
Escalloniaceae (£0.50)	60	84	84
Euphorbiaceae	146	103	83
Ficoidaceae – see Aizoaceae			
Flacourtiaceae (£1.70)	14	132	58
Flagellariaceae (£0.50)	174	14	161
Fumariaceae (£0.50)	8	76	15
Gentianaceae	107	162	134
Geraniaceae (£1.00)	35	92	42
Gesneriaceae	115	176	147
Goodeniaceae (£0.50)	94	187	139
Gramineae	188	9	188
Part 1 (£4.40)			
Part 2 (£6.85)			
Part 3 (£24.50)			
Guttiferae (£1.00)	24	123	76
Haemodoraceae in part – see Agavaceae			
Haloragaceae (£1.00)	65	150	50
Hamamelidaceae (£0.50)	64	87	88
Hernandiaceae (£1.70)	139	74	4
Hippocrateaceae – Celastraceae in part			
Hugoniaceae – Linaceae in part			
Hydnoraceae	133	52	12
Hydrocaryaceae – see Trapaceae			
Hydrocharitaceae	154	8	154
Hydrophyllaceae	108	166	140
Hydrostachyaceae	132	38	27
Hypericaceae (£1.00)	23	124	75
Hypoxidaceae	164	24	181
Icacinaceae (£1.00)	45	110	99
Ilicaceae – see Aquifoliaceae			
Illecebraceae – Caryophyllaceae in part			
Iridaceae	162	29	176
Irvingiaceae – Ixonanthaceae in part			
Ixonanthaceae (£1.40)	32A	93A	82A
Juncaceae (£1.00)	175	20	186
Juncaginaceae (£0.50)	181B	4A	155A
Labiatae	121	169	153
Lauraceae	138	73	3
Lecythidaceae (£0.50)	70	142	71
Leguminosae (£30.00)	57	90	87
Part 1, Mimosoideae (£4.35)			
Part 2, Caesalpinioideae (£5.75)			
Part 3 / Part 4] Papilionoideae (£27.70)			
Lemnaceae (£1.00)	180	13	173
Lentibulariaceae (£1.00)	114	177	146
Liliaceae	169	21	168
Linaceae (£1.00)	31	93	40

	A	B	C
Lobeliaceae (£4.75)	95A	185A	138B
Loganiaceae (£1.20)	106	161	125
Loranthaceae	142	50	104
Lythraceae	72	140	45
Malpighiaceae (£1.00)	33	100	81
Malvaceae	27	117	80
Marantaceae (£1.00)	158	33	167
Melastomataceae (£2.35)	71	147	72
Meliaceae	42	99	112
Melianthaceae (£0.50)	53	112	114
Menispermaceae (£1.00)	4	69	10
Menyanthaceae – Gentianaceae in part			
Mimosaceae – Leguminosae in part			
Molluginaceae – Aizoaceae in part			
Monimiaceae (£0.50)	137	72	2
Montiniaceae (£0.50)	76A	148A	49A
Moraceae	148	42	95
Moringaceae	55	80	18
Musaceae	160	30	164
Myricaceae	150	40	92
Myristicaceae	136	71	5
Myrothamnaceae (£1.00)	63	86	89
Myrsinaceae (£2.10)	99	155	124
Myrtaceae	69	146	70
Najadaceae	182	6	159
Nectaropetalaceae – Erythroxylaceae in part			
Nesogenaceae (£0.85)	120A	168A	152A
Nyctaginaceae	123	57	53
Nymphaeaceae	6	65	8
Ochnaceae	40	121	68
Octoknemaceae – Olacaceae in part			
Oenotheraceae – see Onagraceae			
Olacaceae (£1.00)	44	46	101
Oleaceae (£1.00)	102	160	126
Oliniaceae (£0.50)	73	138	47
Onagraceae (£1.00)	75	149	48
Opiliaceae (£0.50)	46	47	102
Orchidaceae	156	35	185
Part 1, Orchideae (£5.90)			
Part 2, Neottieae, Epidendreae (£12.85)			
Part 3, Vandeae			
Orobanchaceae (£0.50)	113	175	145
Oxalidaceae (£1.00)	36	91	43
Palmae	176	11	179
Pandanaceae	177	2	180
Papaveraceae (£0.50)	7	75	14
Papilionaceae – Leguminosae in part			
Passifloraceae (£1.75)	78	134	62
Pedaliaceae (£1.00)	117	174	149
Periplocaceae – Asclepiadaceae in part			
Phytolaccaceae (£0.50)	128	58	35
Piperaceae	135	37	13
Pittosporaceae (£1.00)	16	85	56
Plantaginaceae (£0.50)	122	179	137
Plumbaginaceae (£1.00)	97	157	136
Podostemaceae	131	44	26
Polygalaceae	17	101	22
Polygonaceae (£1.00)	130	54	33
Pontederiaceae (£0.50)	171	18	170

For Product Safety Concerns and Information please contact our EU
representative GPSR@taylorandfrancis.com Taylor & Francis Verlag GmbH,
Kaufingerstraße 24, 80331 München, Germany

Printed and bound by CPI Group (UK) Ltd, Croydon, CR0 4YY
01/05/2025
01858499-0001